# Eastern Deciduous Forest

WILDLIFE HABITATS

Milton W. Weller, Series Editor

Lowell W. Adams, *Urban Wildlife Habitats: A Landscape Perspective*

Milton W. Weller, *Freshwater Marshes: Ecology and Wildlife Management* (third edition)

Richard H. Yahner, *Eastern Deciduous Forest: Ecology and Wildlife Conservation* (second edition)

# Eastern Deciduous Forest

ECOLOGY AND WILDLIFE CONSERVATION

SECOND EDITION

*Richard H. Yahner*

University of Minnesota Press
Minneapolis
London

Published by the University of Minnesota Press
111 Third Avenue South, Suite 290
Minneapolis, MN 55401-2520
http://www.upress.umn.edu

Library of Congress Cataloging-in-Publication Data

Yahner, Richard H.
Eastern deciduous forest : ecology and wildlife conservation / Richard H. Yahner. — 2nd ed.
p. cm. — (Wildlife habitats)
Includes bibliographical references (p. ) and index.
ISBN 978-0-8166-3360-9 (pb)
1. Forest ecology—East (U.S.) 2. Wildlife conservation—East (U.S.) I. Title. II. Wildlife habitats (Unnumbered)
QH104.5.E37 Y34 2000
577.3′0974—dc21 99-050663

Printed in the United States of America on acid-free paper



27 26 25 24 23 22 21 10 9 8 7 6

*To three special friends:*
*my wife, Darlinda,*
*and my sons, Rich and Tom*

CONTENTS

# PREFACE

More than three centuries ago, European colonists settled the deciduous forest along the eastern coast of the North American continent. These colonists soon became familiar with the forest's animals and plants—resources essential to the colonists' survival—long before venturing westward into prairie and other ecosystems of this vast country. Thus, as a nation, we have deep, historical ties with the eastern deciduous forest.

Today, the eastern deciduous forest is familiar to most Americans. A large percentage of us live within easy driving distance of this forest. To some of us, the forest consists of extensive tracts of "deep woods." To others, it is represented by small urban parklands or isolated woodlots amid extensive agriculture. Many of us have experienced the forest's solitude and beauty while hiking, birding, hunting, or simply sitting beside a quiet woodland stream.

The first edition of *Eastern Deciduous Forest: Ecology and Wildlife Conservation* was published in 1995. I was honored to receive the 1997 Conservation Education Award in the book category from The Wildlife Society for that book. My intent in this second edition remains the same: not merely to describe the eastern deciduous forest or provide a cookbook on "how to manage" for wildlife, but also to give the reader a general introduction to the ecology and wildlife conservation of this magnificent forest. Another goal is to stimulate interest in and encourage a better appreciation of contemporary issues confronting society and natural resource managers who conserve and manage this forest. This book is intended for a general audience: high school students and teachers with environmental interests, beginning college students, and laypeople who are concerned about the environment and love the outdoors.

The second edition's initial chapters deal with basic topics in ecology and wildlife conservation. For the most part, later chapters focus on contemporary conservation issues. Our understanding of the ecology and wildlife conservation associated with the eastern deciduous forest has rapidly developed over the past few years. Since the early to mid-1990s, numerous scientific studies have been published that give us a deeper, and perhaps more sophisticated, knowledge of matters relevant to the eastern deciduous forest, such as forest fragmentation, edges, corridors, biodiversity conservation, and global climatic change. Consequently, I have expanded chapters 3 through 10 to give readers updated information and a better appreciation of important topics. This constantly growing body of knowledge is a modest reminder that we still have a considerable way to go to fully understand the ecology and wildlife conservation of the eastern deciduous forest. I have also provided an extensive reference list of background as well as current literature.

We must be optimistic about the future of the eastern deciduous forest, although many issues will affect the ecology and the conservation of its resources for some time to come. One reason for this optimism is that contemporary issues, such as forest fragmentation and loss of biodiversity, are now being addressed by natural resource managers and the scientific community. For example, numerous natural resource managers, scientists, agencies, and organizations joined forces in 1990 to form the "Partners in Flight Program" to better ensure the future conservation of migratory songbirds in North America. A second reason for optimism is that the public, informed in part by the media and by educators, has recently become more aware of conservation issues. For instance, the worldwide concern about the current accelerated loss of species in tropical forests was brought into our homes by the media at the 1992 Earth Summit in Rio de Janeiro. Our younger generation shares this awareness. My son and his classmates in elementary school during the early 1990s were well informed about the ongoing destruction of tropical forests and its implications for the loss of biodiversity.

As a child growing up in western Pennsylvania, I spent many hours with my father while hunting, walking, or just sitting quietly in the woodlands behind my boyhood home. Later, as a professional wildlife ecologist, I've been fortunate to study a variety of wildlife, ranging from butterflies to white-tailed deer, in the eastern deciduous forest. My research has taken me to surface-mined forests of the mountains in eastern

Tennessee, farmland woodlots in southeastern Ohio, and farmstead shelterbelts of agricultural areas in southern Minnesota. More recently, my studies and those of my graduate students have taken us to the deep "woods" of northern Pennsylvania as well as to woodlands amid agricultural and surface-mined landscapes on private and public lands throughout the state. As a result, I feel comfortable in calling the eastern deciduous forest my home. I hope that my understanding and appreciation of the eastern deciduous forest will help stimulate an interest in the wildlands of the eastern United States. If this book helps to increase the "environmental literacy" of some readers from different walks of life, I will have achieved my goals. Perhaps this book will be an impetus for others to become actively involved in conserving and wisely managing the beautiful eastern deciduous forest and its resources for generations to come.

ACKNOWLEDGMENTS

My research on the eastern deciduous forest has been funded by several agencies and organizations during my career: the Pennsylvania State University, the University of Minnesota, the Pennsylvania Game Commission, the Pennsylvania Wild Resource Conservation Fund, the U.S. Forest Service, the National Park Service, the U.S. Environmental Protection Agency, the Hammermill and International Paper Companies, the Max McGraw Wildlife Foundation, the Hawk Mountain Sanctuary, the Ruffed Grouse Society, the National Rifle Association, the Western Pennsylvania Conservancy, and GPU/Penelec. I am very grateful to personnel of these agencies and organizations for their cooperation and interest in my research, and to my graduate students for their fine research. I thank Rick Sharbaugh for the graphics and Marc Abrams, Barbara Coffin, Jim Lynch, Bill Sharpe, John Skelly, Kim Steiner, and Steve Thorne for helpful reviews of specific chapters in the first edition. I also thank Milt Weller, series editor, for his thoughts and encouragement as I wrote both editions of this book. I would especially like to acknowledge the patience of my family during the time I spent writing this book.

# 1. THE FOREST AND ITS WILDLIFE

## Forests as a Resource

Forests cover about 34 percent of the total land worldwide (Durning 1994). They are found where adequate soil moisture and growing seasons exist, and extend from the tropics to the fringes of the treeless polar tundra. In the contiguous forty-eight states, forests currently comprise about 33 percent of the total land area, or nearly 300 million hectares (Robertson and Gale 1990; Sedjo 1991). In the eastern half of the United States, the percentage of forestland is somewhat higher at approximately 40 percent (Hagenstein 1990).

Although broad in geographic distribution, forests are not an inexhaustible resource. Of the total forestland worldwide, only about 35 percent of it remains relatively unaltered by humans. The rest has been impoverished by excessive deforestation, pollution, or other misuses (Durning 1994). Tropical forests, for example, have been lost at an unprecedented rate in recent decades. Most deforestation in the tropics (e.g., Central America) has occurred since the 1950s, and an estimated 17 million hectares of tropical forests were lost annually between 1980 and 1991 (Laarman and Sedjo 1992; Durning 1994). As we shall see in the next chapter, the amount of forestland in the eastern United States has declined dramatically since European settlement beginning in the seventeenth century, because of farming, logging, and other land uses to meet society's needs.

Forests are an economic and a recreational treasure for all Americans (Cutter, Renwick, and Renwick 1991; Hagenstein 1990; Durning 1994).

Figure 1.1. The eastern deciduous forest contains a diversity of trees and other vegetation and is home to many kinds of wildlife. (Photograph courtesy of the School of Forest Resources, Pennsylvania State University.)

They reduce soil erosion, protect watersheds, absorb precipitation, have a major effect on the stability of the world's climate by releasing oxygen and absorbing carbon dioxide, provide habitat for numerous species of wildlife, and serve as a place of solitude where we can seek relaxation and recreation (Figure 1.1). Furthermore, trees within forests produce at least five thousand products of value to humans (Patton 1992). Trees are a renewable source of wood for home construction, furniture, fuel, raw materials for paper, and many other products common in our homes and workplaces (Laarman and Sedjo 1992). Therefore, a challenge to all of us is to use forest resources wisely without depleting or degrading the forest for wildlife.

## The Eastern Deciduous Forest

A forest can be defined as a group of trees within a given area, but this description is both simplistic and subjective. To a rural New Englander from a small farm at the foothills of the Green Mountains, a "group of trees" may be represented by the large expanses of woodlands covering nearby mountain ridges. Yet to a person from a large city along the East Coast, a forest may be visualized as a woodlot a few city blocks away. To someone from a large Midwestern farm, a forest might be the eight to ten rows of trees in the farmstead shelterbelt planted on the windward side of the nearby farmhouse.

Instead, let us define a forest as either a community or an ecosystem. A forest community includes animals and plants in an area dominated by trees and other woody vegetation (Spurr and Barnes 1980; Hunter 1990). A forest ecosystem encompasses not only these biotic components but also the physical components, such as soil, water, and nutrients. A forest community can be classified arbitrarily by the major or dominant trees in the area, for example, as a beech-maple forest or other forest cover type (Spurr and Barnes 1980). Alternatively, a forest can be termed coniferous (evergreen) when the tree species composition is dominated by pines and related trees, or *deciduous* when trees such as oaks and maples predominate. Occasionally, the terms *softwood* and *hardwood* or *broad-leaved* also are used in reference to coniferous and deciduous trees, respectively (Young and Giese 1990).

Tree species in the eastern United States are principally deciduous, meaning that leaves are shed each autumn (Shelford 1963). Deciduous trees are believed to have evolved as an adaptation to cold winters and occur primarily where growing seasons are relatively long, warm, and humid, as in the eastern United States (Walter 1973). In contrast, coniferous trees predominate in the western United States. The West generally is characterized by harsher growing seasons created by higher elevation and reduced amounts of precipitation compared to the East, except for the mountainous areas along the Pacific coast where rainfall is sufficient. Under the harsher climatic conditions of the West, conifers, particularly in early growth stages, can outcompete deciduous trees (Shelford 1963; Walter 1973; Barnes 1991).

The forest in the eastern United States is separated geographically from that of the West by a relatively nonforested landscape of grassland, remnant prairie, and extensive farmland in the midsection of the country (U.S. Department of Agriculture 1968) (Figure 1.2). The U.S. Department of Agriculture and Society of American Foresters have divided this eastern forest into three broad regions: Lake States and Northeast, Central Mountains and Plateaus, and Southern States (Patton 1992). In eighteen of the thirty-one states falling entirely or partially within one or more of these three regions, more than 50 percent of the land area is forest, ranging from Louisiana with 51 percent to Maine with 90 percent. The forest spanning eastern North America is divided into six ecologically based regions (Figure 1.3). Four of these regions—northern hardwoods, central broad-leaved, southern oak-pine, and bottomland hardwoods—

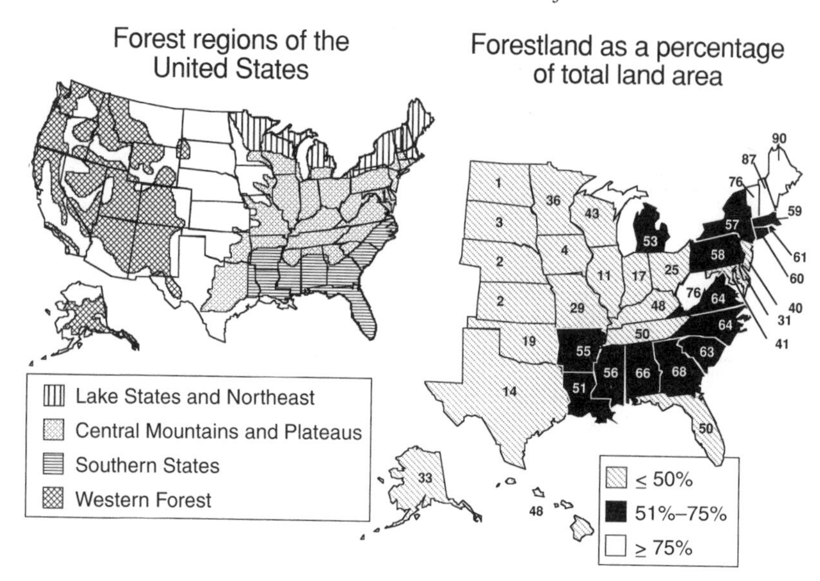


Figure 1.2. Extent of eastern and western forests of the United States and the percentage of forestland per state. The three forest regions in the East are Lake States and Northeast, Central Mountains and Plateaus, and Southern States. Forests in the West are termed Western Forests in this book for simplicity. (Modified from Patton 1992.)

are what we will define in this book as the eastern deciduous forest of North America (Duffield 1990). This magnificent forest is bounded northward in southern Canada by the northern coniferous forest, eastward by the Atlantic coast, southward by the Gulf coast and tropical forest in southern Florida, and westward by grassland, prairie, and extensive agricultural land.

The eastern deciduous forest is composed of an impressive diversity of landscapes and ecosystems that differ to some extent in topography, soil, vegetation, and wildlife (Barnes 1991), and certain forest cover types tend to be dominant in a specific region (Patton 1992). For instance, the aspen-birch cover type is dominant in the northern hardwoods forest region, whereas the oak-hickory cover type is characteristic of the central broad-leaved forest region. Likewise, individual tree species are typically more prevalent in a given forest region (Table 1). Quaking and bigtooth aspen are especially common in northern hardwoods, and green ash and river birch are specific to bottomland hardwoods. On the other hand, some species, such as red maple and northern red oak, are relatively ubiquitous in the eastern deciduous forest.

## Ownership Patterns and Trends in Today's Forest

Forestland ownership in the United States has remained reasonably stable since the 1940s (Hagenstein 1990). About 41.3 percent of the forestland is owned by federal, state, and county agencies (public lands), 45.4 percent by private landowners, and 13.3 percent by private forest and timber companies (forest industry) (Society of American Foresters 1981) (Figure 1.4). Public lands include national forests owned and managed by federal agencies, such as the U.S. Forest Service (Society of American Foresters 1981). The Forest Service owns and manages about two-thirds of this federal land. Other forestland is owned and managed either by other federal agencies, including the National Park Service, the Fish and Wildlife Service, and the Bureau of Land Management, or by state and county agencies (Robertson and Gale 1990).

In the eastern United States, only about 3 percent to 12 percent of the forestland in each state is federally owned; in contrast, as much as 70 percent of the land in each western state is owned by federal agencies,

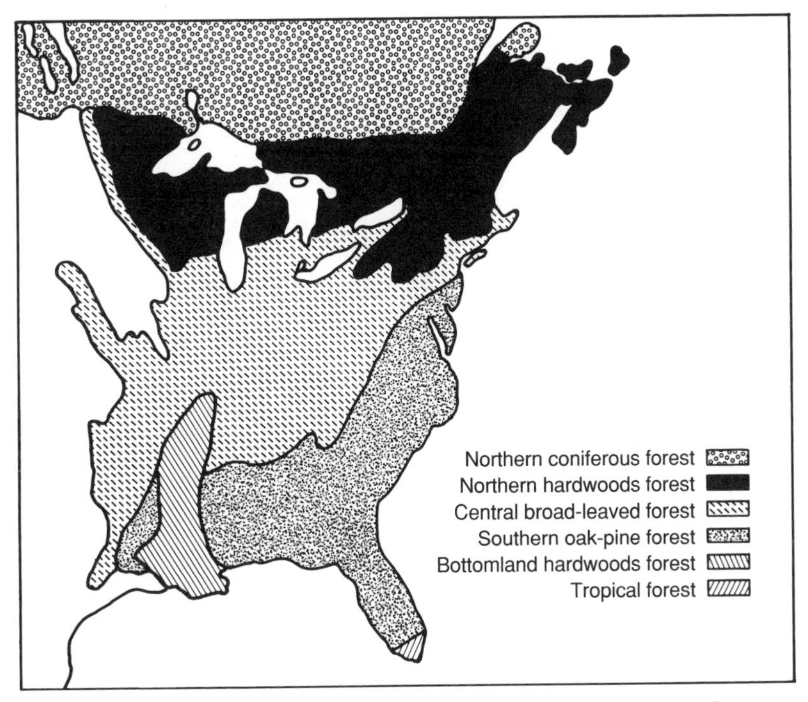


Figure 1.3. The eastern deciduous forest of North America comprises four regions: northern hardwoods, central broad-leaved, southern oak-pine, and bottomland hardwoods. (Modified from Duffield 1990.)

**Table 1. Important tree species (indicated by an X) in four regions of the eastern deciduous forest**

| | Forest region | | | |
|---|---|---|---|---|
| Tree species | Northern Hardwoods | Central Broad-leaved | Southern Oak-Pine | Bottomland Hardwoods |
| Ash, Green | | | | X |
| Ash, White | X | X | | |
| Aspen, Bigtooth | X | | | |
| Aspen, Quaking | X | | | |
| Basswood, American | X | X | | |
| Basswood, White | | X | | |
| Beech, American | X | X | | |
| Birch, River | | | | X |
| Birch, Yellow | X | X | | |
| Buckeye, Yellow | | X | | |
| Cedar, Atlantic White | | | | X |
| Cedar, Eastern Red | | | X | |
| Cedar, Northern White | X | | | |
| Cherry, Black | X | | | |
| Cottonwood, Eastern | | | | X |
| Cottonwood, Swamp | | | | X |
| Cypress, Bald | | | | X |
| Dogwood, Flowering | | X | X | |
| Elder, Box | | | | X |
| Elm, American | X | X | | |
| Hemlock, Eastern | X | X | | |
| Hickory, Bitternut | | X | X | X |
| Hickory, Mockernut | | X | X | |
| Hickory, Shagbark | | X | X | |
| Locust, Black | | X | | |
| Locust, Honey | | | | X |
| Magnolia, Cucumber | | X | | |
| Magnolia, Southern | | | | X |
| Maple, Red | X | X | X | X |
| Maple, Silver | | | | X |
| Maple, Sugar | X | X | | |
| Oak, Black | | X | X | |
| Oak, Blackjack | | X | X | |
| Oak, Bur | | X | | |
| Oak, Cherrybark | | | | X |
| Oak, Chestnut | | X | | |
| Oak, Live | | | | X |
| Oak, Northern Red | X | X | X | |
| Oak, Northern Pin | | | | X |
| Oak, Overcup | | | | X |

| Tree species | Forest region | | | |
|---|---|---|---|---|
| | Northern Hardwoods | Central Broad-leaved | Southern Oak-Pine | Bottomland Hardwoods |
| Oak, Post | | X | | |
| Oak, Scarlet | | | X | |
| Oak, Southern Red | | | X | |
| Oak, Swamp Chestnut | | | | X |
| Oak, Water | | | X | |
| Oak, White | | X | X | |
| Oak, Willow | | | X | |
| Pecan | | | | X |
| Persimmon, Common | | X | | |
| Pine, Jack | X | | | |
| Pine, Loblolly | | | X | |
| Pine, Longleaf | | | X | |
| Pine, Pond | | | | X |
| Pine, Red | X | | | |
| Pine, Sand | | | X | |
| Pine, Shortleaf | | | X | |
| Pine, Slash | | | X | |
| Pine, Virginia | | | X | |
| Pine, White | X | X | | |
| Poplar, Tulip | | X | X | |
| Spruce, Red | X | | | |
| Sugarberry | | | | X |
| Sweetgum | | X | X | X |
| Sycamore, American | | | | X |
| Tupelo, Black | | X | X | |
| Tupelo, Swamp | | | | X |
| Tupelo,Water | | | | X |
| Walnut, Black | | X | | |

*Source*: Duffield 1990.

primarily by the U.S. Forest Service and the Bureau of Land Management. No other nation in the world, with the exception of Costa Rica (Boza 1993), has set aside such a large percentage of its land for its citizens' use and enjoyment as national forests (a total of 155); major national parks (50); national battlefields, memorials, monuments, and sites (306); and national wildlife refuges (452) (Miller 1992). Of these federal lands, 32 percent of the national forests, 12 percent of the major national parks, 49 percent of the national battlefields, monuments, and sites, and 27 percent of the national wildlife refuges are within the eastern deciduous forest.

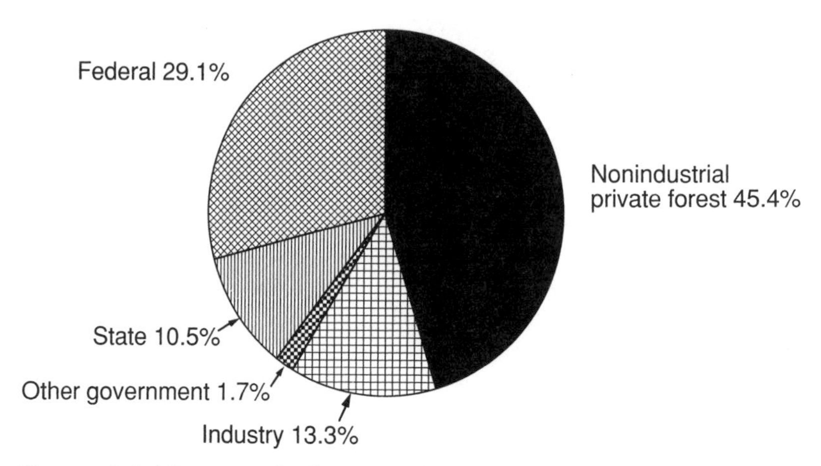


Figure 1.4. Public, private landowner, and private industrial ownership of forestland in the United States. (Modified from Society of American Foresters 1981.)

Forestland in the United States is classified by foresters into three categories depending on whether it is managed specifically for timber: noncommercial forestland, reserved timberland, and commercial forestland (Cutter, Renwick, and Renwick 1991; Hagenstein 1990). Noncommercial forestland, which does not economically produce timber, represents 29 percent of the total forestland. Only about 10 percent of this noncommercial forest is in the eastern United States (Guildin 1990). Since 1952, 16 percent of the noncommercial forestland in the contiguous forty-eight states has been lost to alternate land uses, such as urban development and livestock pasture. About 5 percent of the forestland is reserved timberland, which is in parks and wilderness areas; these lands are off-limits to timber harvesting and are more common in the western United States. Acreage in reserved timberland has tripled nationwide since 1952, largely by the designation of federal wilderness areas, and currently totals about 13.1 million hectares (Bonnicksen 1990).

About 66 percent of the forestland in the contiguous forty-eight states is commercial and considered harvestable, but this percentage differs from East to West (Hagenstein 1990). Roughly 95 percent of the total forestland in the eastern United States is commercial forestland versus less than 50 percent in the West. Commercial forestland is owned mainly by individuals on farms and small landholdings (57 percent), and to a lesser extent by public agencies (28 percent) and the forest industry (15 percent) (Martin and Bliss 1990; Cutter, Renwick, and Renwick 1991).

In the East, the private forest industry (e.g., Georgia Pacific, International Paper, and Weyerhaeuser) has major holdings in Florida, Georgia, Maine, South Carolina, and in some of the Great Lakes states. Private forestland also is owned and managed as farms and landholdings by approximately 7.8 million individual landowners, and most of these holdings are each less than 40 hectares (Martin and Bliss 1990). This type of private ownership is the principal pattern throughout most of the remaining states within the eastern deciduous forest (Cutter, Renwick, and Renwick 1991). In Pennsylvania, for example, 75 percent of the forestland is owned by nearly one-half million private forest landowners. Thus, forest-management practices and their effects on wildlife in the eastern deciduous forest fall largely in the hands of the private forest landowner.

A significant, recent trend in forests of the United States over the past several decades is that slightly more timber is being grown than is being cut (U.S. Department of Agriculture 1982; Cutter, Renwick, and Renwick 1991). More timber is present because much of today's forest is second-growth (Miller 1990); trees on second-growth and younger stands grow more rapidly compared to trees on older stands. The amount of forestland also has increased in the United States. Commercial forestland, for instance, has expanded from about 187 million hectares in 1944 to 196 million hectares in 1977, and the amount of noncommercial forestland has grown from 51 million hectares in 1930 to approximately 103 million hectares in 1977 (Williams 1992). In some areas of the East, the amount of forestland in general has increased dramatically, particularly in New England (from 47 percent in 1880 to 87 percent in 1980), in large part because of conversion of farms to forest, a trend that began in the late nineteenth century (Litvaitis 1993).

Despite growth of timber production in some areas of the country, overcutting of forests has occurred in the United States. A case in point is the remaining old-growth forests in the West. Here timber companies have essentially depleted the amount of harvestable timber on their properties and thus have become more reliant on federal lands where limited amounts of old-growth forests still exist. To many, these old-growth forests are unique and irreplaceable ecosystems and cannot be equated to second-growth forests in terms of their value, for example, to certain wildlife species that require large, undisturbed tracts of forest.

This loss of old-growth forest and its value to wildlife will be discussed in more detail later in this book.

## Wildlife in Today's Forest

Like the generic term *forest, wildlife* has no universal definition (Caughley and Sinclair 1994; Bolen and Robinson 1995). In the past, the term *wildlife* referred to game species, such as white-tailed deer and eastern wild turkey. Revenues raised by the sale of hunting licenses and taxes on firearms and ammunition were earmarked largely for habitat management of these important game species. Fish were treated differently from wildlife as early as 1940, with the merger of the Bureau of Biological Survey and the Bureau of Fisheries to form the U.S. Fish and Wildlife Service. This definition of wildlife continued until at least the 1970s, when wildlife research and management began to focus not only on game species but also on nongame species, such as songbirds and small mammals (Zagata 1978; Hunter 1990).

The definition of wildlife has been expanded by some to include animals and plants except those under the direct control of humans, such as domesticated animals and cultivated plants (Harris and Silva-Lopez 1992; Bolen and Robinson 1995). Some wildlife biologists and society may find this liberal definition difficult to accept in the presence of a double label for the U.S. Fish and Wildlife Service, and given that wildlife journals (e.g., *Journal of Wildlife Management*) continue to publish scientific articles confined mainly to birds and mammals, and that universities and colleges routinely offer degree programs designated specifically as "wildlife" or "fish." The expanded definition of wildlife also becomes problematic as some domestic species become "wildlife." Well-known examples of domesticated species that have crossed the line between domestication and wildness are horses, which form large feral herds in western states (Slade and Godfrey 1982), and pigs, which have become a serious pest in some areas of the eastern deciduous forest (Sweeney and Sweeney 1982).

A liberal definition of wildlife is appropriate, however, today and in the discussion of the eastern deciduous forest in this book. Songbirds, butterflies, and numerous other organisms are appealing "wildlife" to the growing body of outdoor enthusiasts who use and know the eastern deciduous forest. Increasingly, monies are becoming available to state and federal wildlife agencies via voluntary tax checkoffs or various taxes that provide for research and management of nontraditional wildlife species, such as threatened species of plants and nongame animals, and their critical habitats (Hunter 1990). Some scientists even suggest replacing the term *wildlife* with biodiversity (Brussard, Murphy, and Noss 1992; Perry 1993). The term *biodiversity* is sometimes used to refer to the variety of life that exists in our natural world.

About 750 native and naturalized species of trees occur in forestlands throughout the continental United States (Little 1979). These lands are inhabited by approximately 320 species of forest amphibians, reptiles, birds, and mammals (Patton 1992). The number of invertebrate species, which include insects, spiders, mollusks, and other lesser-known life-forms, may easily reach well in the tens or perhaps hundreds of thousands. Invertebrates are important food for many vertebrates (Spurr and Barnes 1980; Mastrota, Yahner, and Storm 1989; Kitchings and Walton 1991) and serve to break down leaves and woody material on the forest floor through the action of bacteria and fungi, thereby releasing important nutrients for uptake by roots of trees and other forest vegetation (Spurr and Barnes 1980).

When we think of wildlife in the eastern deciduous forest, species that readily come to mind are white-tailed deer, gray squirrels, ruffed grouse, eastern wild turkeys, songbirds, and even eastern box turtles (Figures 1.5 and 1.6). The forest contains more than 110 species of trees

Figure 1.5. White-tailed deer in the eastern deciduous forest. (Photograph by Timothy Kimmel, School of Forest Resources, Pennsylvania State University.)

Figure 1.6. Eastern box turtle in the eastern deciduous forest. (Photograph courtesy of the School of Forest Resources, Pennsylvania State University.)

(see Table 1), of which about 75 percent are deciduous (Little 1971, 1979), as well as hundreds of wildflowers and other herbaceous and woody species growing on the forest floor (Gleason and Cronquist 1963). The eastern deciduous forest also provides habitat for many invertebrate species and at least two hundred species of vertebrates. Seventeen species of amphibians (e.g., salamanders, toads, and frogs) occur in the forest (Patton 1992). The southern Appalachian Mountains, in particular, contain an impressive diversity of woodland salamander species (family Plethodontidae) found nowhere else in the world (Kitchings and Walton 1991). Twenty-three species of reptiles (e.g., snakes, lizards, and turtles) inhabit the eastern deciduous forest, most being found in the southern two-thirds (Patton 1992).

In comparison to amphibians and reptiles, many more species of birds and mammals inhabit the eastern deciduous forest. Their higher numbers are not surprising, because bird and mammal species collectively outnumber amphibian and reptilian species in the United States by nearly three to one (1,556 versus 590 species) (U.S. Congress 1987). About 154 bird species are found in the eastern deciduous forest (Patton 1992). More species occur in the northern latitudes because many migratory warblers, such as black-throated green warblers, nest in northern forests where insects abound in the breeding season. In addition, permanent residents, such as ruffed grouse and northern goshawks, extend their geographic range well northward into Canada but are absent from the warmer southeastern states. We could add to this list of bird species by including house sparrows, European starlings, and other exotic species that use rural and urban woodlots in the eastern and midwestern United States (Yahner 1983b).

Birds use various components of the eastern deciduous forest: Woodpeckers and nuthatches forage along trunks and large branches and chip away at the bark of trees in search of food; warblers and vireos forage for insects on small branches and leaves in the forest canopy; flycatchers and tanagers "sit and wait" for flying insects in open areas of the forest; wood thrushes, eastern towhees, ovenbirds, and some warblers feed on or near ground level; and hawks and vultures soar within and over forests searching for food (Kitchings and Walton 1991).

Approximately forty-three species of mammals occupy the eastern deciduous forest, excluding exotic mammals, such as Norway rats and house mice (Patton 1992). Like the number of bird species, the number of mammalian species is higher in the northern two-thirds of the eastern states. Most mammals in the eastern deciduous forest are rodents (mice, voles, squirrels) and insectivores (shrews, moles), which generally forage on the forest floor, and bats, which forage in and above the forest canopy (Kitchings and Walton 1991). Larger mammals common to the eastern deciduous forest are striped skunks, raccoons, red and gray foxes, white-tailed deer, and black bears.

In summary, the eastern deciduous forest is more than just a "group of trees." Rather, it is a complex assemblage of fauna and flora whose beauty and diversity rival those of many ecosystems throughout the world. Today's eastern deciduous forest, however, is a product of thousands of years of changes, some natural and others human-induced, in particular by land-use changes since European settlement. As I point out throughout this book, many human-induced changes have unfortunately been detrimental, or perhaps loom as potentially detrimental, to the conservation of resources in the eastern deciduous forest.

## Wildlife Management—Populations, Habitats, and People

Before the 1930s, wildlife management largely involved the legal protection of birds and mammals and the enforcement of game laws (Bolen and Robinson 1995). The field of wildlife management began to emerge as a science in 1933, with the publication of the classic textbook *Game*

*Management* by Aldo Leopold, the father of wildlife management. Concurrently, the 1937 Federal Aid in Wildlife Restoration Act was passed, which provided a financial basis for states to develop and expand programs for wildlife management.

Through the 1960s, wildlife management dealt almost exclusively with game species (Bolen and Robinson 1995). Wildlife management, however, is much more than the regulation of populations. The wildlife biologist now must also deal with habitats and people (Giles 1978; Bolen and Robinson 1995). In some cases, habitats must be manipulated for the conservation of wildlife or perhaps simply preserved to eliminate human disturbance. People must be educated about the values of wildlife and the importance of rational management and conservation actions. The wildlife manager now must understand what the public thinks about wildlife issues, particularly those that are controversial.

Wildlife management today also relies on sound scientific studies to address problems and issues. Hence, the wildlife biologist is often called upon to be both a manager and a researcher (Bolen and Robinson 1995). Wildlife biologists must use the results of scientific research to remain on the cutting edge of wildlife management and to solve the many conservation problems and issues that affect wildlife in a shrinking world. Furthermore, as we will see later in this book, problems and issues that confront managers of the eastern deciduous forest and its resources will not always be best addressed solely by traditional natural resource disciplines, such as wildlife and forest management. Instead, these traditional disciplines must continue to embrace the relatively new, interdisciplinary, and mission-oriented discipline called conservation biology to better meet society's concerns for the sound conservation of the eastern deciduous forest. In this book, I address some relevant conservation problems and issues that affect wildlife in the eastern deciduous forest, starting with some that began with European settlement and moving on to those that challenge us today and probably will well into the future.

# 2. EARLY HISTORY OF THE FOREST

## Origin of Modern Tree Species

The earliest trees to evolve were the conifers, whose ancestors can be traced back to about 280 million years ago in the geologic era known as the late Paleozoic (Spurr and Barnes 1980). Flowering plants, on the other hand, which eventually led to modern deciduous trees, evolved at least 125 million years ago in the geologic period known as the Cretaceous. Within the next fifteen million years or so after the Cretaceous, flowering plants rapidly dominated the earth. Simultaneously with the arrival of flowering plants, many seed-eating, seedling-eating, and pollinating animals, such as insects and ancestors of modern-day birds and mammals, evolved and spread over the earth.

Ancestors of modern species of trees in the eastern deciduous forest, however, did not evolve until somewhat later, in the Tertiary period, which began about sixty-five million years ago (Spurr and Barnes 1980). North America was virtually covered by deciduous forest by the early Tertiary (Webb 1977). By the late Miocene, beginning about fifteen million years ago, many tree genera had evolved that are still present in today's eastern deciduous forest, such as beech and maple (Spurr and Barnes 1980). As these modern genera of trees were evolving, grasslands began to spread throughout the Great Plains, replacing what was once forest (Webb 1977). During the late Miocene, numerous genera of birds and mammals, for example, songbirds and deer, evolved that still occur in present-day forests of the eastern deciduous forest (Vaughan 1986; Welty and Baptista 1988).

## Ice-Age Effects on the Eastern Deciduous Forest

A little over two million years ago, in the Pleistocene period, major climatic changes were occurring in North America. Glaciers advanced and retreated perhaps sixteen to twenty different times during the Pleistocene, each advance lasting between 50,000 to 100,000 years (Spurr and Barnes 1980; Barnes 1991). As much as 32 percent of the world was covered by ice during the Pleistocene, compared to about 10 percent today (Spurr and Barnes 1980). With each glacial-interglacial cycle in the Pleistocene, the climate alternated between periods of cooling and warming, causing the southern edge of the glaciers to advance and retreat. Because of these north-to-south shifts in glaciers, the species composition and the distribution patterns of trees in the eastern deciduous forest were in a state of flux. About eighteen thousand years ago, during one of these advances known as the Late Wisconsin, the climate cooled sufficiently to enable colder-climate species, such as spruce and jack pine, to extend their geographic range well into the southern United States (Figure 2.1). Spruce invaded areas to the west of the Mississippi River, and jack pine penetrated to the east of the Mississippi River as far south as northern Mississippi, northern Georgia, and South Carolina (Delcourt and Delcourt 1981; Barnes 1991). Warmer-climate species, such as oaks, hickories, and southern pines, were kept below this spruce-jack pine boundary. During warmer interglacial periods, each spanning between ten thousand and twenty thousand years, the ice masses melted at the southerly fringes, allowing the climate to moderate south of the retreating glaciers (Spurr and Barnes 1980). In turn, conditions became favorable for warmer-climate species of deciduous trees to recolonize these sites previously occupied by colder-climate species and glaciers.

Despite persistent changes in the climate throughout the Pleistocene, many genera of deciduous trees continued to exist in North America while going extinct elsewhere. A good example are the hickories, which are found today in the eastern deciduous forest but disappeared from Europe and western Asia during the Pleistocene. Perhaps hickories and other genera long gone from the Eastern Hemisphere survived glaciation in North America because our mountain chains are oriented north to south, allowing distributions of deciduous trees to shift with the coming and going of the glaciers. In contrast, the mountain chains of Europe and western Asia are positioned east to west, prohibiting the retreat and recolonization of deciduous trees as glaciers moved north and south across the Northern Hemisphere (Spurr and Barnes 1980).

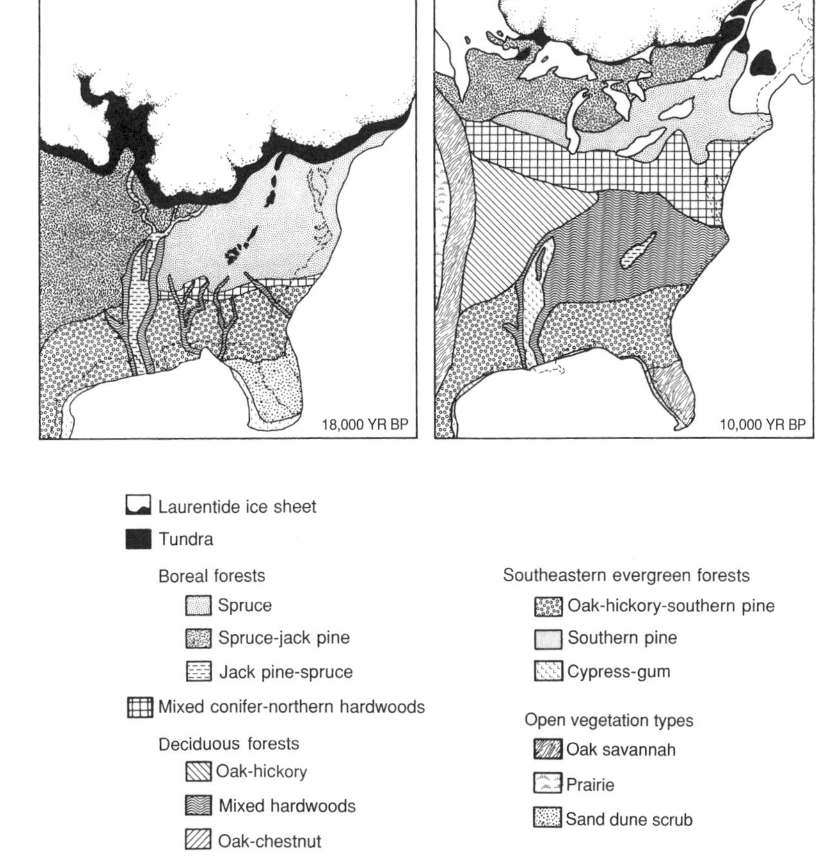


Figure 2.1. Maps of vegetation in eastern North America at eighteen thousand years B.P. (before present) and ten thousand years B.P. (Modified from Delcourt and Delcourt 1981.)

An important warming trend occurred about 16,500 years ago, which marked the final northward retreat of conifers in the southern latitudes of the eastern United States (Barnes 1991). In the wake of this final retreat, only isolated stands of conifers remained at high elevations in the Appalachian Mountains; stands of conifers can still be enjoyed today by hikers in certain areas, such as in the Great Smoky Mountains National Park.

Beginning about 12,500 years ago, deciduous tree species expanded rapidly northward and eastward into what is now the eastern deciduous forest (Figure 2.2). As a consequence of this final glacial retreat, populations of certain wildlife species were left behind in isolated pockets of habitat at high elevations. For instance, black-capped chickadees and northern flying squirrels, which are more typical of northern latitudes in North America, can be found today as remnant populations at higher elevations in the southern Appalachians.

As the glaciers waned after the Pleistocene, the climate of eastern North America entered a major warming trend, which began about ten thousand years ago and is often referred to as the Holocene. Temperatures reached a maximum between five thousand and eight thousand years ago (Barnes 1991), eliminating the last vestiges of ice in the eastern deciduous forest about eight thousand years ago (Spurr and Barnes 1980). During the transition from the Pleistocene to the Holocene, several kinds of magnificent animals became extinct in North America, such as woolly mammoths and mastodons, and large carnivores, such as

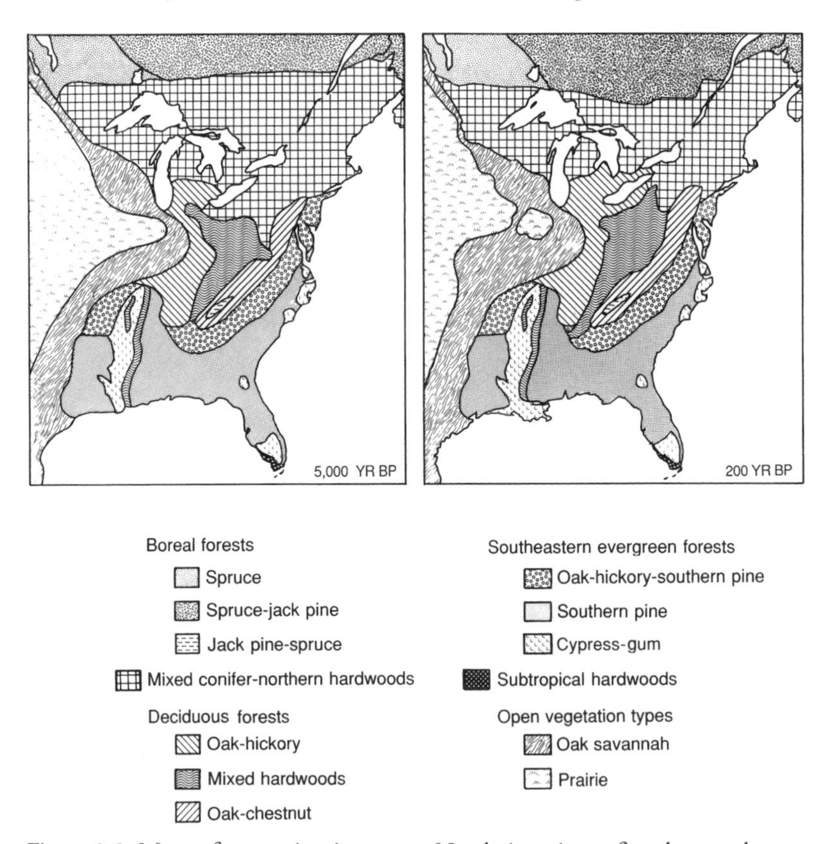


Figure 2.2. Maps of vegetation in eastern North America at five thousand years B.P. and two hundred years B.P. (Modified from Delcourt and Delcourt 1981.)

saber-toothed cats and dire wolves (Matthiessen 1978). Hence, climatic changes over the past several thousands of years have influenced the extinction patterns of many species of forest wildlife. These changes also affected the distribution and the abundance of extant wildlife species in the eastern deciduous forest.

## Pre-European Forests in the Eastern United States

About two-thirds of North America was covered with old-growth forest when Europeans first set foot on the continent in the early seventeenth century (Cutter, Renwick, and Renwick 1991). In the continental United States (including Alaska), about 50 percent, or 445 million hectares, was forested; three-fourths of this forestland was located in the eastern United States (Harrington 1991; MacCleery 1992). The forest of eastern North America remained relatively undisturbed until the late eighteenth century. Early descriptions of the New England forest suggest that the tree species occurring nearly four centuries ago were the same that grow in the region today. White pine, beech, maple, and hemlock, for instance, were among the abundant trees growing in the eastern deciduous forest before European settlement (Braun 1950) (Figure 2.3).

In pre-European times, the eastern deciduous forest was affected by natural events, such as wildfires and windthrows. These events influenced the distribution and the abundance of forest animals and plants encountered by early European settlers (MacCleery 1992). Native Americans are also thought to have had an impact on the pre-European forest, particularly through their use of fire and their agriculture (Spurr and Barnes 1980; MacCleery 1992; Williams 1992; Delcourt and Delcourt 1997). These people carefully used fire to manage the land and promote their livelihood (Williams 1992). In Wisconsin, for instance, at least 50 percent of the forest may have been influenced by fires set intentionally by Native Americans (Curtis 1959). Fires set by Native Americans may have reduced the distribution of some fire-intolerant tree species, such as yellow birch, red maple, and black cherry, while benefiting chestnut oak, white oak, and other fire-tolerant species in the eastern deciduous forest (Day 1953; Nowacki and Abrams 1992; Delcourt and Delcourt 1997).

Most Native Americans lived in villages throughout the northeastern United States (Day 1953; Shelford 1963; Cronon 1983). Each village included at least several hectares of clearings that contained homesites. These clearings were expanded to obtain timber for building homes;

Figure 2.3. Old-growth hemlock forest in Potter County, Pennsylvania. (Photograph courtesy of the School of Forest Resources, Pennsylvania State University.)

making utensils, canoes, and other items; and particularly for use as a fuel. As fuelwood was depleted, villages were relocated. In some instances, larger areas were cleared for agriculture, which was the major means of subsistence of most Native Americans in the eastern deciduous forest (Williams 1992). For example, the Seneca Indians in northwestern Pennsylvania planted hundreds of hectares in corn. Native Americans converted thousands of forested hectares near Massachusetts Bay to agricultural land, probably by using fire. The Iroquoian and other tribes relied greatly on agriculture to provide foodstuffs, such as beans, maize (corn), squash, and sunflowers, grown in forest clearings created by fires that were set deliberately or occurred naturally (Figure 2.4). Fires also were used selectively by Native Americans to create forest edges and openings for deer, turkey, and other wildlife, which were important as food.

## Forest Resource Use by Early European Settlers

For the European settlers arriving in the New World, the original eastern deciduous forest provided an ample source of timber for use as fuel and building materials (Cutter, Renwick, and Renwick 1991). Approximately two-thirds of the wood cut in colonial times was used to warm homes (MacCleery 1992). Clearing of the original eastern deciduous forest began before 1650 in New England and the mid-Atlantic states, eventually converting much of the forest to cropland and pasture (Sedjo 1991) (Figure 2.5). In central Massachusetts, virtually all the original forest was cut and cleared for hay, crops, and pasture by Europeans before the end of the eighteenth century (Spurr and Barnes 1980; Williams 1992). In Pennsylvania, an estimated 4 million of 11.7 million hectares of the original forest were transformed to other uses, principally agriculture, from 1660 to the 1970s (Powell and Considine 1982). Conversion of forest to agricultural use in the South began around 1750 (Sedjo 1991). Forest in the Georgia Piedmont, for instance, was converted to extensive cotton fields in the decades before the Civil War.

In the seventeenth century, wood from New England was vital to the shipbuilding industry of the region (Perlin 1989). New England became a leader in providing the world with fish, whale meat, and whale products. Whaling, especially in pursuit of the species known as the right whale, attributed to the tremendous development of the New England coastline by colonists. Whaling was so intense that by the end of the

Figure 2.4. Native American village showing clearings for crop production and home sites. (From MacCleary 1992.)

Figure 2.5. Farm in Massachusetts at height of agricultural development, circa 1830. (Photograph courtesy of Harvard Forest Models, Fisher Museum, Harvard Forest, Harvard University, Petersham, Massachusetts.)

seventeenth century, the population of right whales was decimated along the New England coasts (Reeves and Brownell 1982).

Early in the seventeenth century, wood from the northeastern United States was an important export product (MacCleery 1992). A considerable amount of lumber was exported to England to provide masts and timbers for shipbuilding. In the later part of the seventeenth century, wood from New England was critical to sugar plantation owners in the British West Indies, where forests had already been devastated (Perlin 1989). Without this wood, which was used to construct and repair sugar works, to make staves for casks to package and export sugar, and to build homes, sugar producers would have shut down the mills. In turn, England and much of Europe would have lost 70 percent of their source of sugar.

Wood throughout the United States was very important as a material to build farm fences until barbed and woven wire were manufactured in the later part of the nineteenth century (MacCleery 1992). By 1850, for example, an estimated 5.1 million kilometers of wooden fences were erected, which is equivalent to more than one thousand coast-to-coast round-trips across the United States. Local and regional shortages of timber began to occur by the mid-nineteenth century in the northeastern states. The center of timber harvesting had shifted by the 1850s from Maine and New York to Pennsylvania, and by later in the century

Figure 2.6. Abandoned farm naturally reforested to pines and later harvested in 1909. (Photograph courtesy of Harvard Forest Models, Fisher Museum, Harvard Forest, Harvard University, Petersham, Massachusetts.)

to the Great Lakes states (Powell and Considine 1982; Cutter, Renwick, and Renwick 1991).

Forests continued to decline in extent in the early twentieth century as agriculture became more widespread and moved westward (Clawson 1979). By the 1850s, at least 47 million hectares had been converted to agriculture in the eastern United States (Williams 1989). With a greater focus on agriculture in the midwestern states, clearing in the East had slowed by the 1920s (Cutter, Renwick, and Renwick 1991; Sedjo 1991). Many farms were abandoned along and east of the Appalachians, allowing land used previously as farmland to revert back to forest. Abandoned farmland often reverted to pines; when left unburned, these pine stands developed dense understories of deciduous trees (Figure 2.6). As a result of this conversion of farmland to forest, forestland in the forty-eight contiguous states has increased an estimated 20 percent since the early twentieth century (Cutter, Renwick, and Renwick 1991). Forests cut over in the early twentieth century also were being replaced with second-growth stands, which provided valuable wildlife habitat and protected important watersheds (Powell and Considine 1982).

Agriculture and fuelwood cutting were not the only major impacts on the eastern deciduous forest. In many areas, forests were repeatedly grazed by cattle or cut for use in iron production (Spurr and Barnes 1980; Marquis 1983). Forests of Pennsylvania and West Virginia, for example, were clear-cut regularly to provide fuel for iron production and railroad engines. The mining industry used a considerable quantity of wood for timbers, and charcoal was used to fuel the iron furnaces (Figure 2.7). During the eighteenth century until the late nineteenth century, virtually all the iron produced in the United States was smelted using wood charcoal (MacCleery 1992). Charcoal operations were a major cause of forest fires in Pennsylvania during the late nineteenth and early twentieth centuries (Ward 1983).

Various land uses eliminated some species, such as eastern hemlock and white pine, from many areas in the eastern deciduous forest, especially in the northern hardwood and central broad-leaved forest regions (Duffield 1990) (see Figure 1.3). Tons of hemlock bark were used by the tanning industry. Intensive logging and fires restricted the distribution of white pine, whereas most oak species proliferated because of their ability to regenerate from stump sprouts (Nowacki and Abrams 1992).

Diseases also affected the distribution and the abundance of tree species in the eastern deciduous forest subsequent to European settlement. Chestnut blight and Dutch elm disease were introduced by settlers and nearly eliminated certain tree species (Spurr and Barnes 1980; Patton 1990). The fungus causing chestnut blight arrived from Asia in 1904 and eradicated most American chestnuts in New England within twenty years after introduction. By the 1940s, the American chestnut was almost extirpated from the southern Appalachians, except for occasional

Figure 2.7. Charcoal making in 1900. (Courtesy of the Forest History Society.)

Figure 2.8. The American beaver of North America. Dams created by beaver modify and provide habitat for many species of wildlife, such as waterfowl and other aquatic mammals. (Photograph by Robert P. Brooks, School of Forest Resources, Pennsylvania State University.)

root sprouts that seldom survived long enough to grow into mature trees. The fungus responsible for Dutch elm disease, introduced into Ohio from Europe around 1930, nearly wiped out American elms from mesic and lowland forests throughout the East and continues to be a problem. As American chestnuts and American elms were lost to disease in the eastern deciduous forest, they were replaced by other dominant tree species, such as black cherry and red maple.

## Wildlife in the Original Forest

Wildlife was abundant in the eastern deciduous forest before European settlement. Many species of wildlife were, and continue to be, important to the teachings and the culture of Native Americans (Storm 1972). Europeans arriving in the New World wrote that bears, turkeys, and fish abounded in forests and streams (Cronon 1983). They also had some interesting descriptions of wildlife. Moose in New England, for instance, were described as being "of the bigness of a great house" (Thomas Morton, as quoted in Perlin 1989, 272). The ruby-throated hummingbird fascinated the English colonists because "as she flies, she makes a little humming noise like a Humble bee: wherefore she is called the humbird" (William Wood, as quoted in Perlin 1989, 273).

Early colonists were impressed by the American beaver's (Figure 2.8) "art and industry" (William Wood, as quoted in Perlin 1989, 273). Because of its fur, valued throughout the world, the beaver was the most sought-after natural resource in the eastern deciduous forest during the 1700s and 1800s (Hill 1982). Beaver trapping quickly attracted European trappers to areas west and north of the eastern forest. A beaver population that perhaps exceeded sixty million in North America was virtually extirpated by the 1900s. Livetrapping and restocking of the beaver as well as natural dispersal in the mid-1900s, however, resulted in its return to much of the eastern deciduous forest. In many parts of the beaver's range, it has now become so common that it occasionally causes damage to trout streams and timber, particularly where beaver dams cause extensive flooding in flat terrain (Hill 1982).

As the eastern deciduous forest was modified by European settlers via farm clearings, timber harvesting, and wildfires, many other species of wildlife became less abundant (MacCleery 1992). Game species, such as turkey, elk, and bison, were scarce or absent in many areas by the end of

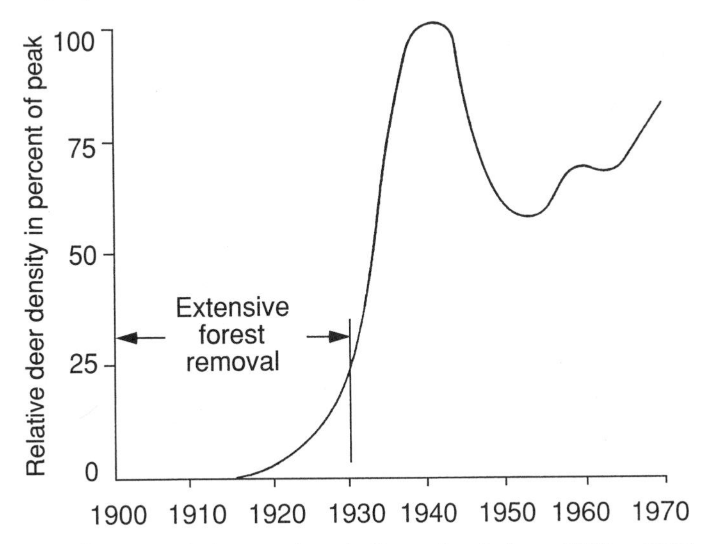


Figure 2.9. Deer population numbers in Pennsylvania from 1900 to 1970. (Modified from Marquis 1975.)

the eighteenth century (Harrington 1991). Even songbirds, such as American robins, were hunted as food. Passenger pigeons and heath hens had been extirpated from the East by the turn of the twentieth century (Bolen and Robinson 1995).

In contrast, populations of white-tailed deer in pre-European times in the eastern deciduous forest were probably higher than today (Mattfield 1984; McCabe and McCabe 1984; Nowak 1999). Deer had been nearly eliminated by the early 1900s because of unregulated hunting (Shrauder 1984). A combination of state hunting regulations, live-trapping and restocking, lack of natural predators, and extensive logging in the early twentieth century created favorable habitat for deer, and populations have increased dramatically in the eastern deciduous forest (Figure 2.9). In Pennsylvania, for instance, the deer population was estimated at approximately one million in the late 1930s (Forbes et al. 1971). The white-tailed deer, almost annihilated at one time from the eastern deciduous forest, attracts more attention from hunters, outdoor enthusiasts, wildlife professionals, and society in general than any other wildlife species. But, as I shall discuss later, problems resulting from "too many" deer loom as a major wildlife and forest management issue in the eastern deciduous forest.

# 3. ECOLOGICAL PROCESSES

## Ecology—The Study of Interrelationships

Ecology is defined as the study of the natural environment and of the relations of organisms, including plants, animals, microbes, and people, to each other and to their surroundings (Ricklefs 1990). The term *ecology* is derived from *oikos,* which is Greek for "household," and from *logg,* which means "the study of." Ecology is a complex science that seeks to understand patterns of the distribution and the abundance of organisms. Hence, forest ecology deals with the interrelationships among organisms within the forest community as well as the physical environment in which these organisms live (Spurr and Barnes 1980). Ecology is concerned not only with the individual organism, such as a tree or bird in relation to its environment, but also with groups of organisms, populations, communities, ecosystems, and even landscapes. Animal ecology has often been treated separately from plant ecology (Krebs 1972), but the distribution and the abundance of many animals are dependent on plants and vice versa; thus, an understanding of both is vital to a full appreciation of the complex nature of the eastern deciduous forest.

Numerous abiotic and biotic factors determine where and how many plants and animals may occur in a forest. The influence of such factors differs between plants and animals because individual plants are stationary, whereas animals generally are mobile. A good example of this differential effect is that of soil moisture on forest plants versus animals. If a drought occurs in a localized area within the eastern deciduous forest, a mature maple tree cannot escape the harsh, dry conditions on the hillside in which it is rooted, but a broad-winged hawk can move to an area with more suitable conditions for nesting.

Abiotic and biotic factors, such as solar radiation and herbivory, respectively, can also profoundly influence the regrowth of forests, a process termed forest succession or regeneration. The concept of forest succession as well as specific plant and animal interactions, for example, herbivory, will be given expanded attention in the next two chapters. In this chapter, I present a brief overview of several important abiotic and biotic factors that are part of the natural environment and that influence relations of organisms in the eastern deciduous forest.

## Abiotic Factors in the Eastern Deciduous Forest

### *Solar Radiation*

Solar radiation, or sunlight, is vital to life in the forest. Solar radiation enables plants to transform solar energy into chemical energy via the process of photosynthesis (Ricklefs 1990). Photosynthesis combines carbon dioxide ($CO_2$) and water ($H_2O$) and releases oxygen ($O_2$), forming the carbohydrate glucose ($C_6H_{12}O_6$) (Figure 3.1). This process gives plants the energy and the carbohydrate building blocks to produce tissues for growth and reproduction. Plants in turn are a source of food (leaves, fruit, etc.) on which forest animals depend. The accumulation of energy and nutrients by plants via photosynthesis is termed net primary production. Net primary production in the eastern deciduous forest is lower than that in tropical forests and freshwater marshes but higher than that in boreal forests and grasslands (Whittaker and Likens 1973).

The amount of solar radiation received by the eastern deciduous forest varies with a number of factors, such as season, time of day, and direction of slope (Spurr and Barnes 1980). Greater radiation occurs in summer than in winter. Regardless of season, radiation is greater at midday than at sunrise and sunset and greater on south-facing than on north-facing slopes. The distribution and the abundance of many plants and animals in the eastern deciduous forest are influenced by these differences in solar radiation. Because the sun's rays strike more directly on south-facing slopes, conditions are warmer as well as dryer than on

$$6C0_2 + 12H_2O \rightarrow C_6H_{12}O_6 + 6O_2 + 6H_2O$$

Figure 3.1. The chemical equation for photosynthesis.

north-facing slopes; hence, vegetation more typical of drier, warmer conditions generally prevails on southerly slopes (Collier et al. 1973). Nests of red-tailed hawks more often occur in trees on slopes facing the southeast; this exposure maximizes the amount of solar radiation reaching the nest on cold mornings and reduces heat stress in the late afternoon (Speiser and Bosakowski 1988). Tree cavities and nest boxes used by nesting American kestrels also tend to face a southerly or an easterly direction (Brauning 1983; Rohrbaugh and Yahner 1997) (Figure 3.2). Similarly, Appalachian (eastern) woodrats, whose current geographic range extends only as far north as Pennsylvania, use homesites (e.g., rock outcrops) on warmer, south-facing slopes more often than on cooler, north-facing slopes in the mountains of Pennsylvania (Balcom and Yahner 1996).

During winter, about 50 to 80 percent of the solar radiation reaches the floor in the leafless eastern deciduous forest, whereas only about 1 to 5 percent penetrates the canopy in summer (Spurr and Barnes 1980). Moreover, in winter, when deciduous trees are devoid of leaves and solar radiation passes easily through the canopy, ambient temperatures within a forest stand may be a few degrees higher than outside the stand.

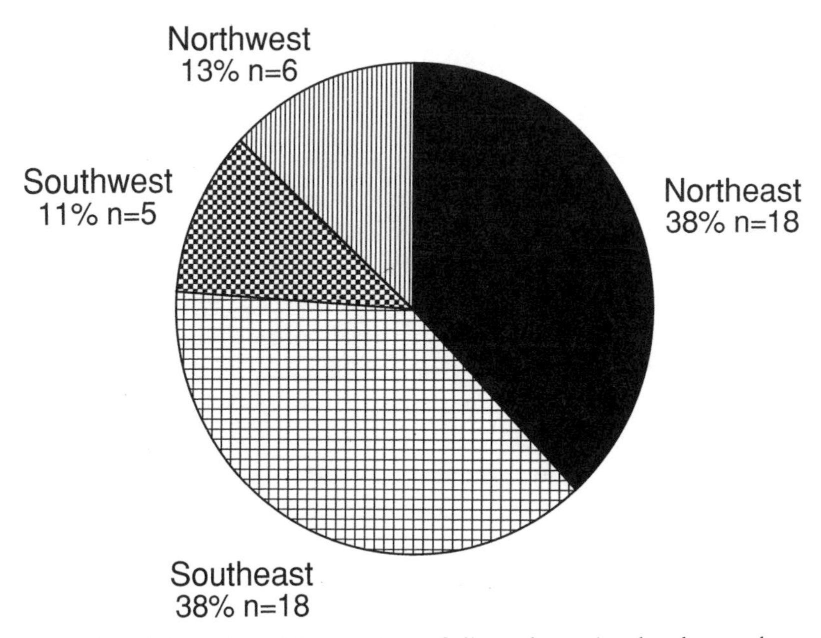


Figure 3.2. Orientation of the openings of all nest boxes (used and unused versus frequently used nest boxes) by American kestrels in Pennsylvania.

Conversely, air temperatures in a forest stand in summer with full leaf-out may be cooler than outside the stand. Colder temperatures and reduced day length in winter eliminate plant growth, and plants become dormant over winter. The forest comes back to life in the spring as day length and ambient temperatures increase. The period between snowmelt and leaf-out of eastern deciduous trees is extremely important to successful germination and growth of ephemeral herbaceous plants, such as the beautiful violets that cover the forest floor in early spring (Muller 1978). Violets reduce the loss of important nutrients, particularly potassium and nitrogen, in the forest soils by accumulating them and later returning them to the soil as the plants die in summer (Muller and Bormann 1976).

When the canopy of a forest stand becomes more open because of a natural event (e.g., a tornado) or a human-induced event (e.g., logging), a greater amount of solar radiation reaches the ground level. This radiation in turn warms the soil and facilitates the lush growth of shade-intolerant plant species, such as grasses, forbs, and trees (e.g., aspen), which need considerable sunlight in early growth stages (Miller 1990). A dense layer of vegetation benefits numerous species of forest wildlife; for example, it provides abundant *browse* for white-tailed deer, nest sites for common yellowthroats, and cover for southern red-backed voles (Yahner 1992, 1993a).

### *Ambient Temperature*

Ambient temperatures typically decrease in the eastern deciduous forest as latitude and elevation increase. Because mountain chains along eastern North America run parallel to the Atlantic coast, temperatures decline along a relatively uniform gradient with increasing latitude northward into Canada and from the lower elevations along the Atlantic coast to the crest of the Appalachians (Spurr and Barnes 1980). Ambient temperatures, such as solar radiation, influence the geographic distribution of forest plants and animals. Oaks in the southerly latitudes of the eastern deciduous forest, for instance, are replaced with northern hardwoods, which are better adapted to colder temperatures in more northerly latitudes (see Figure 1.3, Table 1). The distribution of animals, such as eastern cottontails and Virginia opossums, also is limited by winter cold. Eastern cottontails have adapted to many habitat conditions from forest to urban in the eastern United States, but they exhibit considerable weight loss and, hence, intolerance of winter temperatures in northern states (Chapman, Hockman, and Edwards 1982). In contrast, the New England cottontail, although limited in distribution to the northern New England states, is well suited to the colder winter temperatures. The Virginia opossum has expanded its range southward in the eastern deciduous forest over recent decades, yet the northerly expansion of this marsupial probably has terminated in southern Canada because of its susceptibility to frostbite and severe winters (Gardner 1982).

Ambient temperatures can affect seasonal and daily activities of forest animals. The timing of emergence from winter dormancy in forest mammals, such as eastern chipmunks and black bears (Figure 3.3), is triggered in part by the spring warming trend and the availability of spring foods (Lindzey and Meslow 1976; Yahner and Svendsen 1978). On the other hand, emergence into fall hibernation by some animals, such as pregnant black bears, likely occurs only when sufficient body fat reserves have been accumulated to ensure winter survival, pregnancy, and lactation (Schooley et al. 1994).

Many songbirds, like the scarlet tanager, escape winter conditions in the eastern deciduous forest by migrating in late summer or autumn to Central and South America (Terbough 1989). Some northern bats, such as the eastern red bat, likewise migrate to more southerly latitudes. The

Figure 3.3. The black bear is a common forest mammal in the northeastern United States. (Photograph courtesy of the School of Forest Resources, Pennsylvania State University.)

big brown bat, which is often referred to as the "streetlight" bat because it forages around streetlights at dusk, does not migrate but instead hibernates in warm caves over winter (Humphrey 1982). Brown bats probably use heated buildings as winter hibernacula more often than previously suspected (Whitaker and Gummer 1992). As many as eighty-six bats were found overwintering in heated homes and churches with insulated attics. Hibernating bats likely have gone undetected in buildings during winter because they are few in number, seldom make noise, and do not defecate. Incidentally, buildings are also important as sites for bat maternity colonies in spring and summer. Bat boxes constructed and erected by people concerned about bat conservation have become increasingly valuable as substitute maternity sites for bats displaced from buildings (Williams and Brittingham 1997).

American beaver are active outside the water and lodge throughout most of the year, but in northern latitudes during winter, the animals confine most activity to the lodge, where temperatures can be at least 20° to 30°C warmer than ambient temperatures (Dyck and MacArthur 1993). In western states, black bears of cinnamon, brown, or other non-black color phases are occasionally active in open meadows during the midday in summer (Rogers 1980). In contrast, western black bears exhibiting the black color phase, which is the typical color phase of eastern black bears, are rarely active in open habitats at these times. One proposed explanation for avoidance of the midday summer sun by black-phased black bears in the West is that they are more susceptible to heat stress than black bears of a paler color. The western black-phased bear instead avoids high ambient temperatures and direct solar radiation by foraging during cooler mornings and evenings.

Ambient temperatures can affect homesite selection by wildlife. In the Southeast, black bears often hibernate in tree cavities averaging 11 meters aboveground (range = 5–20 meters) (Johnson and Pelton 1981). The placement of a den aboveground in a tree cavity provides insulation from the winter cold, thereby reducing fat loss by about 15 percent. Dens located in tree cavities also protect bears from predators or flooding caused by heavy winter rains. Thus, large den trees (e.g., ≥ 84 centimeters diameter at breast height) can be an important limiting resource to bear populations in southerly latitudes. The preservation of large trees with cavities is of particular interest to wildlife biologists involved with the restoration of bear populations into historic ranges of the South (Oli, Jacobson, and Leopold 1997). Bears in the southern states occasionally den at ground level, but only in elevated areas where dense vegetation gives protection from both predators and floods (Wooding and Hardisky 1992). Conversely, black bears in more northerly latitudes generally den at or just below ground level in protected sites, such as rock cavities, hollow tree stumps, and brushpiles (Alt and Gruttadauria 1984). Snow-covered den sites in the North not only conceal a hibernating bear from predators, but dens also save the bear energy by reducing fat loss by about 27 percent. Warm dens and reduced fat loss better enable a pregnant or nursing female bear to raise her young. For this reason, any disturbance of a hibernating black bear with young should be avoided, because it can have a dramatic negative effect on reproductive success (Tietje and Ruff 1980).

For hibernation in winter, striped skunks also seek sheltered dens, which usually are within abandoned burrows or under buildings (Gunson and Bjorge 1979). Unlike bear dens, however, skunk dens are often occupied by one male and an average of six females. Communal denning increases the temperature within the den about 2° to 3°C for each additional animal, thereby saving energy stored as body fat. Moreover, communal denning by skunks ensures successful breeding, which occurs in early spring.

White-tailed deer in northern parts of their range congregate in sheltered areas, termed deeryards, to conserve energy in winter (Rongstad and Tester 1969). In more southern latitudes, deer instead remain relatively sedentary throughout the year but may make limited movements to coves or hillsides with dense vegetative cover, such as mountain laurel, to stay warm in winter (Schilling 1938; Storm, Cottam, and Yahner 1995).

### *Soil Texture and Layers*

Soil is the substrate within which most plants in the eastern deciduous forest grow. Soil particles contain organic materials (decaying leaves, etc.) and minerals, which include important nutrients for plant growth. Water and air spaces are found in the pores between soil particles. Soil texture is based on the size of soil particles, and soils can be classified into sandy, silty, or clay soils (Spurr and Barnes 1980). Sandy soils have particles from 0.2 to 2.0 millimeters in diameter, silty soils have particles from 0.02 to 0.002 millimeters in diameter, and clay soils have particles less than 0.002 millimeters in diameter. Sandy soils hold less water and fewer minerals than clay soils; in contrast, clay soils contain less air than sandy soils. Soils with approximately equal amounts of sand, silt, and clay are referred to as loams.

Soil texture can affect plant and animal distribution. Certain tree species, such as red pine and scrub oak, do well in sandy soils but less well than in soils of different texture and of higher nutrient content (Spurr and Barnes 1980; Bockheim 1990). In contrast, loamy soil is required by many tree species in the eastern deciduous forest, for example, red maple, black cherry, and tulip poplar. Loamy soil is ideal for forest mammals that burrow, such as the eastern chipmunk, which relies on extensive burrow systems for home and hibernation sites (Svendsen and Yahner 1979).

We can separate the surface soil, or organic (O) horizon, of the eastern deciduous forest into an upper litter layer, containing unaltered remains of dead plants and animals; a fermentation layer, with altered remains of plants (leaves, twigs, fruit, bark) and animals that still can be identified; and a humification layer, consisting of altered remains that are not identifiable (Spurr and Barnes 1980) (Figure 3.4). The humifica-

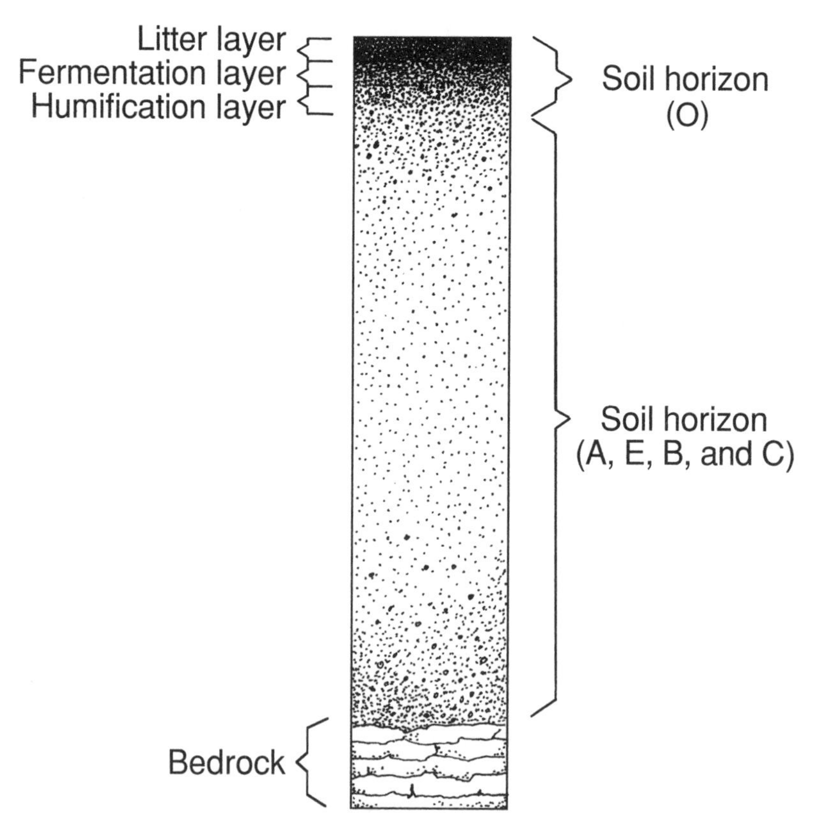


Figure 3.4. Soil layers in the eastern deciduous forest. (Modified from Spurr and Barnes 1980.)

tion layer is broken down to minerals, carbon dioxide, water, and humus. Below the O horizon are four additional horizons (A, E, B, and C), which except for the upper A horizon lack organic matter (Bockheim 1990).

Roots of most forest trees are often within 2 to 3 meters of the soil surface; over 90 percent of small roots (< 2.5 millimeters) of pines and oaks, for instance, are in the upper 13 centimeters of the surface soil (Coile 1937; Kozlowski 1971). Burrowing forest animals also stay relatively near the soil surface. Winter dens of black bears and burrow systems of woodchucks, for example, are typically less than 1 meter deep in the forest soil (Grizzell 1955; Alt and Gruttadauria 1984; Wooding and Hardisky 1992).

Leaf litter in the surface soil is decomposed by bacteria and fungi (Spurr and Barnes 1980). Forests with well-drained soils typically have faster rates of litter decomposition and, hence, a thinner layer of leaf litter than forests with poorly drained soils. In some mature forests, as much as 1,500 to 5,000 kilograms per hectare of leaf litter may accumulate per year. Leaf litter is especially important to many species of forest-floor wildlife, such as shrews and eastern towhees, which search the leaf litter for invertebrates for food (Merritt 1987; Mastrota, Yahner, and Storm 1989; Gross 1992).

### *Soil Nutrients*

Soil nutrients are vital to the growth of plants. A ready source of nutrients is organic matter—freshly fallen leaves, twigs, stems, bark, and flowers—which fall to the forest floor, decompose, and are recycled in the eastern deciduous forest (Aber 1990; Bockheim 1990). Conversely, nutrients are lost from the forest as trees are removed for timber and other forest products (Lorimer 1990b). Deciduous trees in the eastern forest typically require more nutrients for growth than conifers, which are often capable of thriving on nutrient-poor soils. The amount of nutrients needed by deciduous or coniferous trees is much less, however, than that required by most lawns in suburbia.

Nutrient uptake by forest trees is different from uptake by potted houseplants. Houseplants have numerous small root hairs that help accumulate nutrients and water. Roots of forest trees, on the other hand, do not have abundant root hairs; instead, they have small fungi, known as mycorrhizae, that grow on rootlets (Spurr and Barnes 1980; Kimmerer 1990). More than four hundred fungi species are known to form mycorrhizae in North American trees (Marx and Beattie 1977). The mycorrhizal fungi, which are present in small amounts in the soil, give the rootlets a swollen appearance. These fungi seem to help trees accumulate nutrients, especially phosphorus and nitrogen, and water in a mutually beneficial relationship by increasing the surface area of the roots (Harley 1969). In turn, nutrients and water become available for use by these fungi. Mycorrhizae are especially essential to the growth of trees, such as maples, that are located on sites with low soil nutrients or during periods of drought. Mycorrhizae on roots also may reduce disease and insect damage (Zak 1965).

Nitrogen and calcium are two nutrients very important to tree growth in the eastern deciduous forest (Spurr and Barnes 1980). Most nitrogen in the forest soil or atmosphere is not usable by trees (Kimmerer 1990). Some eastern deciduous trees, for example, black locust and alder, that grow in nitrogen-poor soil can, however, convert atmospheric nitrogen to organic nitrogen, which then can be taken up readily by the rootlets. This nitrogen conversion is made possible by the growth of microorganisms, such as specific types of bacteria, on rootlets of these trees.

Certain trees, such as flowering dogwood and tulip poplar, concentrate large amounts of calcium, which was obtained from the soil, in their foliage; other trees, such as American beech and eastern hemlock, are low in calcium uptake (Spurr and Barnes 1980). Calcium is a major nutrient affecting the acidity of forest soil. The pH of forest soil can range from 4.0 to 7.5 and tends to be less acidic (pH = 7.0–7.5) in calcium-rich soils.

Animals of the eastern deciduous forest also require nutrients for normal growth. An excellent example is the white-tailed deer, which requires small amounts of calcium and phosphorus for proper antler and skeletal development (French et al. 1956; Ullrey et al. 1973). Low levels of soil calcium possibly are linked to declines in egg production and reductions in the geographic distribution of upland game birds in the eastern United States, most notably of the ring-necked pheasant (Dale 1954, 1955).

### *Soil-Water Relationships*

The major source of water in the eastern deciduous forest is precipitation in the form of rain or snow that percolates into the soil (Spurr and Barnes 1980). Not all precipitation, however, reaches the forest floor; some hits the foliage in the canopy and eventually is evaporated back into the atmosphere. More precipitation is intercepted by the foliage in a deciduous forest than in a coniferous forest, and snowfall is intercepted better by the foliage than rainfall is (DeWalle 1995). In some instances, as much as 70 percent of the total precipitation can be recycled into the atmosphere from the forest, and the amount of water entering the forest soil is lessened further by surface runoff (Spurr and Barnes 1980). In a typical year, 40 to 60 percent of the annual precipitation may return as streamflow in the eastern deciduous forest (DeWalle 1995). The forest soil is usually very wet and at full water capacity near the beginning of the growing season (Spurr and Barnes 1980). Soil moisture becomes depleted but is replenished by rainfall as the growing season progresses. In general, an average of 600 and 1000 millimeters of rainfall is needed for forest tree growth during the growing season in northern and southern states, respectively.

Soil water is critical to nutrient cycling in the forest ecosystem (Spurr and Barnes 1980; Bockheim 1990). As plants absorb water through root systems, nutrients suspended in the water are transported in the plant. Uptake of water is necessary for photosynthesis, as we saw earlier. Furthermore, in order for trees to grow properly, there must be an optimal balance between the amount of water and air space in the soil. If there is too much water, thereby leaving inadequate air space in the soil for oxygen, trees may "suffocate," because oxygen is vital for plant respiration. Too much water is especially harmful to trees adapted to upland sites in the eastern deciduous forest, such as some oaks and hickories. In addition, excessive water in the soil, for example, during flooded conditions, can become a problem by removing the supply of soil nutrients needed for tree growth. Thus, when upland sites are flooded for a prolonged time period, root mortality, reduced tree growth, and even tree death may be the outcome. Conversely, very wet or flooded conditions usually have negligible or no effect on growth and survival of swamp and river floodplain species, such as cottonwoods and willows.

Moisture in the forest soil can vary depending on the degree of slope and the position on the slope, which in turn can influence the distribution of plants and animals. Sites with steep slopes in the central broadleaved region of the eastern deciduous forest (see Figure 1.3), for instance, usually have relatively dry soils compared to sites lower down the slope; hence, chestnut oak predominates on upper, dryer sites, whereas tulip poplar is abundant on lower, moister sites (Gleason and Cronquist 1963). In this same forest region, forest small mammals and songbirds have higher population densities and a greater diversity of species on more mesic slopes at middle to low elevations compared to upper, dryer slopes (Yahner and Smith 1990, 1991).

Periodic moderate droughts can occur during the growing season, resulting in reduced soil moisture. Most trees in the eastern deciduous forest typically can withstand these moisture shortages (Spurr and Barnes 1980). But shortages of water can affect the distribution and the abundance of forest insects that depend on moist habitat conditions; these changes may in turn affect populations of vertebrates, such as songbirds, that depend on these organisms for food (Petit, Petit, and Grubb 1985). The drought of 1988 caused a marked decline in localized populations of the masked shrew, which is a forest small mammal that is dependent on both moist microenvironments and invertebrate food in the leaf litter (Yahner 1992).

## Biotic Factors in the Eastern Deciduous Forest

The eastern deciduous forest consists of an assemblage of plant and animal species whose distribution and abundance are not only a function of abiotic factors but, to some degree, of biotic factors. Two principal biotic factors that deserve some discussion in this chapter are competition and predation.

### *Competition*

Competition may occur when two individuals of the same or different species coexist within a given forest stand and require the same resource. Competition among forest plant species for limited solar radiation, soil nutrients, and moisture can reduce growth or survival of individual plants. As trees in a recently harvested forested stand become larger and older, competition can increase among tree species. For example, in the northern hardwood forest region, northern red oaks continually expand their canopy crown, thereby outcompeting coexisting white ash for solar radiation, nutrients, and moisture (Spurr and Barnes 1980). Many herbaceous plants, shrubs, and small trees beneath the forest canopy also compete for vital resources. In recently cut forest stands, rapid growth of blackberry can eliminate grasses and ferns in northern hardwoods of Pennsylvania (Marquis and Grisez 1978).

Forest animals often compete within or among species for limited resources, such as homesites or food. Natural cavities are a limiting resource for cavity-nesting birds in some forest stands. Red-bellied woodpeckers excavate tree cavities for nesting, but this woodpecker is less aggressive than some other woodpeckers, such as red-headed woodpeckers or northern flickers, toward European starlings (Ingold 1994). As a result, nearly 40 percent of red-bellied woodpecker nests in east-central Ohio were usurped by starlings compared to 15 percent or less of nests of the other two woodpecker species. Nest boxes have been shown to reduce competition among nesting birds where natural cavities are rare or absent (McComb and Noble 1981; Yahner 1983/84).

Interspecific competition between two species also can occur as vandalism. The house wren removes eggs and destroys nests of many bird species (Belles-Isles and Picman 1986). This wren has been a major factor causing the decline in the Appalachian Bewick's wren, which once was relatively common in the eastern United States until about the 1930s (Kennedy and White 1996).

Interspecific competition for homesites potentially may occur between sympatric populations of mammals, as suggested in gray and fox squirrels. However, close inspection of homesite selection by these two squirrels has shown that there is very little overlap in use of tree cavities or leaf nests (Edwards and Guynn 1995). Gray squirrels more readily use cavities than fox squirrels do, and leaf nests of gray squirrels are constructed closer to ground level and in smaller trees than nests of fox squirrels. Interspecific competition may exist between fox squirrels and raccoons for tree cavities as homesites (Robb et al. 1996). Tree cavities selected by fox squirrels have smaller entrances and are located farther from water than cavities chosen by raccoons. The smaller entrances of fox squirrel cavities probably reduce the likelihood of predation on squirrel nests by raccoons.

Sympatric birds, such as red-tailed hawks and great horned owls, potentially can compete for food. A recent study documented that although both species are generalist predators, there is about a 50 percent overlap in prey species in the diet of these two raptors (Marti and Kochert 1995). Furthermore, reptiles were an important food type for red-tailed hawks, whereas invertebrates were much more important to the great horned owl diet. Similarly, snowshoe hares and white-tailed deer occupying young clear-cut stands in north-central Pennsylvania rely extensively on browse (woody vegetation) as a food resource (Scott 1986; Scott and Yahner 1989). However, competition between these two mammals may be reduced somewhat because hares feed heavily on browse of blackberry, striped maple, and yellow birch; on the other hand, deer also rely extensively on browse of both red and sugar maple.

Competition in plants may cause direct harm to another organism

(Krebs 1972). Some plants produce and release chemicals into the immediate environment that can negatively affect other species; this action is termed allelopathy (Spurr and Barnes 1980). Chemicals from black walnuts, for instance, prevent the growth of other tree species that come in contact with their roots (Brooks 1951). In the eastern deciduous forest, the growth of yellow birch may be restricted by the presence of sugar maple. Other tree species suspected of causing allelopathic effects in the eastern deciduous forest are black cherry and sassafras (Aber 1990). Grasses and ferns on the forest floor can inhibit the germination and growth of black cherry seedlings in a similar manner (Horsley 1977).

Territoriality is a good example of competition in animals involving aggression among individuals, usually of the same species (Krebs 1972; Caughley and Sinclair 1994). This type of competition is prevalent in forest songbirds during the breeding season: Males use songs, visual displays, and occasional aggressive encounters to defend territories from others. This behavior helps to attract mates and ensures adequate food resources for one or both parents to raise the young (Gill 1990). The ruffed grouse, an aesthetic upland bird species of the eastern deciduous forest, also defends a territory. Here, the male grouse uses an elaborate auditory and visual display, known as drumming, to attract a mate and to ward off rival males (Craven 1989) (Figure 3.5). Unlike many songbirds, however, male grouse do not share in parental duties; the female grouse and her brood usually are not associated with the male's territory.

In general, territoriality is less common in mammals than in birds of the eastern deciduous forest. The red fox, which typically is monogamous and mates for life, defends a territory around a den site; both sexes share parental duties (Storm et al. 1976). The eastern chipmunk also is territorial, but it is solitary, and only the mother cares for the young. Each sex, however, defends a territory centered around the individual's burrow system (Yahner 1978). In contrast to these mammals, the gray squirrel is not territorial but instead lives in a loose "colony" in which individual squirrels know one another, presumably by olfaction; immigration by young animals into the colony is restricted by aggression in adult females (Thompson 1978). The white-tailed deer, like virtually all deer species, is not territorial. There is evidence, however, that an adult female white-tail will defend a limited amount of area around her fawns, called a fawning ground, for a few weeks after their birth (Ozoga, Verme, and Bienz 1982). Defense of a fawning ground probably acts to better ensure adequate food resources for the adult female with young, particularly when deer densities are high.

Figure 3.5. The drumming of a ruffed grouse is used to defend a territory and attract a mate. (Photograph courtesy of the School of Forest Resources, Pennsylvania State University.)

Territoriality in wide-ranging carnivores acts to space individuals or groups according to the amount of available habitat. For instance, based on an intensive long-term study in Minnesota, territories of gray wolf packs had an average radius of 6.4 kilometers (Mech 1994). An interesting finding from this study was that the boundaries of some pack territories were defended aggressively against intrusion by neighboring packs and resulted in significant intraspecific mortality. Of twenty-two gray wolves killed by conspecifics, 91 percent were within a 3.2–kilometer buffer zone to each side of their respective territorial boundary. Perhaps because mortality is higher within these buffer zones compared to more central locations within territories, wolves minimize the amount of time spent in buffer zones (Peters and Mech 1975). This is the first report of mammals killing conspecifics along territorial boundaries, but future research may show that this behavior occurs in other territorial carnivores.

### *Predation*

Predation is a second biotic factor influencing the distribution and the abundance of organisms in the eastern deciduous forest. Some predators at the top of the food chain, such as great horned owls, are strictly

Figure 3.6. Energy transfer from plants to top carnivores in the eastern deciduous forest. (Modified from Ricklefs 1990.)

carnivores, preying on a variety of small vertebrates (Wink, Senner, and Goodrich 1987) (Figure 3.6). Other top predators, such as gray foxes, are omnivores, feeding on both animal and plant foods (Kozicky 1943). Virginia opossums, striped skunks, white-footed mice, and blue jays are additional examples of omnivores in the eastern deciduous forest. Some vertebrates lower on the food chain, including many forest songbirds and small mammals, feed on invertebrates in the canopy, understory, and leaf litter (Merritt 1987; Terbough 1989). Numerous invertebrates, such as forest insects, and vertebrates, such as white-tailed deer, snowshoe hares, and eastern cottontails, are herbivores that rely on plants as food resources. Herbivores at the bottom of the food chain almost always greatly outnumber carnivores at the top of the chain. Rabbits, for example, are much more numerous than foxes in the eastern deciduous forest. The energy transferred from plants to top carnivores in the food chain of the eastern deciduous forest is usually no more than 5 to 20 percent (Ricklefs 1990).

Top predators in the food chain may have a direct influence on populations of smaller predators in the chain. In the northerly latitudes of the eastern deciduous forest, the geographic ranges of three weasel species, the ermine, least weasel, and long-tailed weasel, overlap to some extent (Simms 1979). The southerly distribution of ermines and least weasels may be restricted in part by aggression and potential predation by long-tailed weasels on these two smaller species. Likewise, in areas where the distribution of the smaller red fox overlaps that of the larger coyote, fox territories are limited to habitat unoccupied by coyotes or habitat outside the core areas of coyote territories (Sargeant, Allen, and Hastings 1987; Harrison, Bissonette, and Sherburne 1989). Survivorship of red foxes is higher at boundaries than within core areas of coyote territories because coyotes are typically very aggressive toward foxes and, on occasion, may kill them (Sargeant and Allen 1989).

An interesting by-product of the negative impact of sympatric coyotes on red foxes can be found in the prairie pothole region of South and North Dakota, which is just west of the eastern deciduous forest. This region is critical nesting habitat for a variety of waterfowl, such as mallard and blue-winged teal, and red foxes are major predators on these species (Sargeant 1972). In areas where red foxes were the principal canid, nesting success by waterfowl was 17 percent (Sovada, Sargeant, and Grier 1995). In contrast, in areas where coyotes excluded red foxes, nesting success nearly doubled to 32 percent, presumably because coyotes are less effective predators on waterfowl nests. The management of top predators, like the coyote, in an ecosystem likely will be given more attention in the future because the loss of these predators can lead to extirpations or localized declines in desirable species (Palomares et al. 1995). The decline in songbird populations in some habitats throughout North America also may be the partial result of an abundance of smaller predators whose numbers are no longer held in check because of an absence of larger predators in the ecosystem (Wilcove 1985; Soulé et al. 1988).

Ecologists have been interested in predator-prey relationships for many decades (Vaughan 1986). One such relationship, the snowshoe hare–lynx cycle, is a well-documented population cycle (Keith et al. 1984). This cycle is generally eight to twelve years in duration and occurs in northern latitudes of North America, including the northern hardwoods region of the eastern deciduous forest. The cycle is initiated when browse becomes scarce in winter because of heavy browsing pressure by hares. This food shortage leads to malnutrition in hares and lower population numbers. Hare numbers at the peak of the cycle, for example, may be as high as 7 to 9 individuals per hectare but drop to 0.4 to 1 individuals per hectare at the cycle low (Poole 1994). Lynx, which are very dependent on hares as food, then decline in abundance (Keith et al. 1984). During years of hare scarcity, kitten production ceases and survival of lynx of all ages declines dramatically; furthermore, the proportion of kittens in the lynx population can drop from nearly 50 percent in winters with high hare abundance to zero in winters with low hare abundance (Poole 1994).

Compared to the snowshoe hare–lynx cycle, the vole population cycle is not well understood and remains one of the most famous, unsolved phenomena in mammalian ecology (Lidicker 1988; Krebs 1996). This cycle typically lasts three to four years, and examples of voles that undergo this type of cycle in the eastern deciduous forest are the southern red-backed and the woodland vole. Even though this cycle has been studied since the 1920s, the factor (or factors) causing the vole cycle has not been identified. Currently, some ecologists hypothesize that it is caused by an unknown interaction between both extrinsic (e.g., predators, food availability) and intrinsic (e.g., behavior) factors that may vary with geographic region, year, and perhaps species.

An important question that has been asked concerning predation and predator-prey cycles is whether predators control or regulate numbers or distribution of their prey. Some evidence from European studies suggests that populations of small mammals, such as voles, may be controlled by predators, like weasels and red fox (Erlinge et al. 1984; Lindström et al. 1994). Populations of ungulates, such as deer, elk, and moose, on the other hand, are believed to be regulated principally by food availability rather than by predators (Peek 1980). Although predators, such as gray wolves and mountain lions, may reduce numbers of ungulates, they more often affect their distribution (Hornocker 1970; Mech 1977; Nelson and Mech 1981). For example, an interesting result of high intraspecific mortality of wolves along pack buffer zones, which we discussed earlier, is that survival of white-tailed deer in Minnesota is higher along edges of wolf pack territories than within more centralized areas of territories (Mech 1977). Thus, when populations of white-tailed deer are low, the remaining deer often are restricted to habitat where wolves are less likely to occur while hunting as a means of avoiding fatal encounters with wolves from other packs. Some recent studies, however, surprisingly show that bears are a prominent predator on ungulates, such as moose; 50 percent of moose calf deaths have been attributed to predation by black and brown (grizzly) bears in certain areas of North America (Boutin 1992).

Density of prey can affect the abundance of predators. One of the best documented cases of this effect is the relationship between populations of gray wolves and moose at Isle Royale National Park in Lake Superior (Peterson and Page 1988). Data collected over the past few decades seem to support the contention that wolf survival and pack size are very dependent on the availability of moose, which are the major prey of wolves in the park. As moose numbers decline, survival of individual wolves and pack size are reduced dramatically; ecologists call this phenomenon a numerical response. Alternatively, some predators switch prey items rather than decline in numbers when one type of prey becomes scarce, which is termed a functional response. Great horned owls, for instance, prey extensively on snowshoe hares in northern latitudes; when hares become uncommon, ruffed grouse become primary prey for the owls (Rusch et al. 1972). Interactions between predators and prey clearly leave many interesting and unanswered questions for the wildlife biologist to address in the future.

In summary, we have seen that the ecology of the eastern deciduous forest is very complex. Abiotic and biotic effects on plant and animal distribution and abundance—the "wheres," "hows," and "whys"—are only beginning to be understood, even for the more common or familiar species. We still do not fully appreciate the short- and long-term implications of other important relationships between plant and animal species, such as pollination and seed dispersal. Furthermore, natural resource managers are only beginning to discern the impacts of herbivory by certain forest animals, such as defoliating insects and high populations of white-tailed deer, on resources in the eastern deciduous forest. These plant and animal interactions will be discussed in the next chapter.

# 4. FOREST PLANT AND ANIMAL INTERACTIONS

We have seen that trees, shrubs, spring ephemerals, and other types of vegetation serve many vital functions in the eastern deciduous forest, such as adding oxygen to the atmosphere, absorbing precipitation, and recycling nutrients. Plants are also especially critical to forest animals in two major ways: They provide homesites and food resources. Trees and shrubs provide substrates for nest placement by songbirds, and tree cavities in older or dying trees are resting sites or refugia for small mammals, such as northern flying squirrels and white-footed mice. Berries and nuts are eaten readily by a host of forest birds and mammals, bark of trees is consumed by American beavers and common porcupines (Figure 4.1), and succulent new growth of forbs, leaves, and twigs is important food for deer and other wildlife. Herbivory, however, can have negative consequences for forest plants, particularly when densities of certain forest insects (e.g., gypsy moths) or large mammals (e.g., white-tailed deer) become excessively high. Before examining the adverse impacts of herbivory by select insects and large mammals on the ecology and the conservation of resources in the eastern deciduous forest, I will first discuss two noteworthy ways in which animals benefit plants, that is, as agents of plant pollination and seed dispersal.

## Plant Pollination

Hummingbirds, bats, bees, and other pollinating species of wildlife are crucial to a large number of flowering plants, such as some willows and maples, in the eastern deciduous forest (Giese 1990; Kimmerer 1990).

Figure 4.1. Bark of a small tree removed by a common porcupine. (Photograph by author.)

Without these pollinators, many flowering plants could not reproduce and would not have reached their current distribution and dominance throughout the world (Regal 1977). Many species, such as butterflies, not only act as important pollinators but also are aesthetically pleasing to outdoor enthusiasts in the eastern deciduous forest (Kim 1993). For instance, 129 species of flowering plant species were observed over a three-year period at several forested and agricultural sites in central Pennsylvania (Yahner 1998). Fifty-two (40 percent) of these flowering species were used as a food (nectar) source by twenty-one species of butterflies. Some important wildflowers for butterflies such as great spangled fritillaries, monarchs, and spicebush swallowtails, are common fleabane, goldenrod, heal-all, and thistle.

Today, scientists are becoming increasingly concerned about the conservation of pollinating species in both temperate and tropical regions of the world. We need not look any farther than the northeastern United States, where thousands of honeybee colonies recently have died, with home owners reporting a virtual absence of wild bees in backyards (Finley, Camazine, and Frazier 1996). In winter and spring 1995–96, beekeepers in the Northeast documented significant percentages of colony losses in various states: Delaware, 25–40 percent; Pennsylvania, 53 percent; New Jersey, 60 percent; New York, 60–70 percent; and Maine, 80 percent (Finley, Camazine, and Frazier 1996; Fore 1996). Although a portion of these colony losses were attributed to inclement weather, much of the loss has been blamed on the parasitic mite syndrome (PMS), which is a poorly understood disease in bees (Shimanuki, Calderone, and Knox 1994). Prior to the introduction of PMS, only about 10 percent of bee colonies were normally lost per year (Frazier et al. 1994). This syndrome also is characterized usually by tracheal mites that may stress bees and act as disease vectors, resulting in eventual declines in brood production, lower populations numbers, and colony death (Finley, Camazine, and Frazier 1996). Aggressive treatment for diseases and mites in honeybee colonies has been shown to significantly increase survival and health of honeybee colonies.

Honeybees are vital pollinators for many important crops, such as alfalfa, apples, melons, and pumpkins. The increase in crop yield and value to agriculture in the United States attributed to honeybee pollination is estimated at $9 billion per year (Robinson, Nowogrodzki, and Morse 1989). Within the last few years, losses of honeybee colonies in North America and in some European countries have convinced scientists that management and protection of honeybees and other pollinators are critical to sustaining future human food supplies (Allen-Wardell et al. 1998).

In the prairies of the Midwest, a perennial grass called the royal catchfly is becoming extirpated locally because of a scarcity of pollinators (Menges 1991). Small, isolated populations (< 100 individual plants) of this grass show reduced rates of germination (about 50 percent) compared to larger populations (> 150 individuals, 85 percent germination), in part because small, isolated populations are visited less often by pollinating hummingbirds. In certain regions of the world, such as the tropics, the situation is much more serious. The ongoing extinction of some species of pollinating fruit bats, such as in Guam, will probably lead to the future extinction of many plants dependent on these mammals for reproduction (Cox et al. 1991). Furthermore, a negative by-product of the extirpation of pollinating species in the tropics is the loss of important plant products; about 450 products useful to humans are derived from 289 plant species known to be pollinated by tropical fruit bats (Fujita and Tuttle 1991). Whether an accelerated loss of pollinating animals is likely to occur in the eastern deciduous forest is yet unknown. However, given that this phenomenon is seemingly widespread and that honeybee colonies have declined precipitously, natural resource managers concerned about plant conservation cannot take this issue lightly.

## Seed Dispersal

Plants rely either on wind or animals to disperse seeds and thereby help ensure future germination. The flowering dogwood is a good example of a tree species in the eastern deciduous forest whose seeds are well adapted to fruit-eating birds and mammals (Spurr and Barnes 1980). The dogwood fruit containing the seeds is eaten by animals, but the seeds are undamaged because of a hard outer coating. When excreted by animals, some of the seeds germinate, often at a considerable distance from the parent plant.

Dispersal and subsequent germination of oak acorns in the eastern deciduous forest also are facilitated by animals. Gray squirrels and eastern chipmunks bury individual acorns in the leaf litter on the forest floor throughout their home ranges, a behavior that is called scatter hoarding (Yahner 1977). Many of these scatter-hoarded acorns probably are not recovered, and some successfully germinate into seedlings.

The blue jay is another significant agent of acorn dispersal (Figure 4.2).

Figure 4.2. Blue jay with an acorn.

In one study, blue jays removed, dropped, and scatter hoarded about 54 percent of the acorn crop, totaling 133,000 acorns, in a stand of pin oaks; distances from seed trees to caches averaged 1.1 kilometers (Darley-Hill and Johnson 1981). Because of this ability to transport and scatter hoard large quantities of acorns over considerable distances, some scientists believe that the blue jay was largely responsible for the rapid northward spread of oaks in the eastern United States several thousands of years ago, at the end of the Pleistocene (Johnson and Adkisson 1986).

Dispersal and successful germination of oak acorns are beset, however, by many hazards because nearly 100 species of birds and mammals feed extensively on acorns (Martin, Zimm, and Nelson 1951). Eastern chipmunks, white-footed mice, gray squirrels, white-tailed deer, black bears, and ruffed grouse readily consume acorns in the eastern deciduous forest (Batzli 1977; Merritt 1987; Eller, Wathen, and Pelton 1989; Servello and Kirkpatrick 1989). Home ranges and densities of eastern chipmunks in the eastern deciduous forest, for instance, often are associated with areas containing abundant oaks, because of the nutritional benefit of acorns (Pyare et al. 1993; Lacher and Mares 1996). Population densities and winter survival of gray squirrels are very dependent on favorable production of acorns in the previous autumn (Nixon, McClain, and Donohoe 1975). Year-to-year densities of white-footed mice have been shown to be directly correlated with densities of autumn acorns, which serve as a vital winter food source for this small forest mammal (Elkinton et al. 1996); white-footed mice are known to breed throughout the winter following an autumn with peak acorn production (Ostfeld, Jones, and Wolff 1996). White-tailed deer rely on an autumn supply of acorns as food in order to build up body fat reserves for winter survival (Mautz 1984; Cypher, Yahner, and Cypher 1988) (Figure 4.3).

When deer populations are high in a localized area, they are capable of removing a large percentage of the total acorns within a relatively short time period (McShea and Schwede 1993). This efficient removal of acorns by deer conceivably may limit populations of other wildlife during years of low acorn production. Some evidence suggests that small

Figure 4.3. Evidence of feeding by white-tailed deer on acorns on the snow-covered forest floor beneath a large oak tree. (Photograph by author.)

mammals also are capable of removing a considerable proportion of the available acorns. White-footed mice, for instance, removed (ate or cached) nearly two-thirds of the total acorns during autumn in shelterwood, thinned, and uncut forest stands in central Pennsylvania (DeLong and Yahner 1996). An insect known as the acorn weevil can damage (and hence effectively remove) large quantities (50 percent) of acorns that fall to the forest floor (Andersen and Folk 1993). Interestingly, white-footed mice and northern short-tailed shrews are very efficient at finding and eating weevil larvae that overwinter in the forest soil, thereby having a positive effect on oak regeneration. In addition, gray squirrels can have a major, positive impact on oak regeneration because of their ability to detect acorns infected with weevil larvae (Steele, Hadj-Chikh, and Hazeltine 1996). Gray squirrels immediately eat the infested acorns along with the larvae but cache the noninfested acorns. Thus, cached acorns not retrieved later by squirrels tend to be weevil-free and more likely to germinate.

Certain kinds of acorns contain high concentrations of chemicals called tannins, which make them less palatable to some species of wildlife. Red oak acorns, in particular, have higher contents of tannins than white oak acorns, making red oak acorns less preferred by gray squirrels (Lewis 1982). Tannins apparently are not effective deterrents to all wildlife; white-footed mice readily consume both white and red oak acorns (Briggs and Smith 1989).

In summary, successful pollination and seed dispersal by animals are critical to the ecology of many plants in the eastern deciduous forest. Conversely, excessive loss of seeds, such as acorns, to certain species of wildlife may have short-term effects on other species of wildlife and perhaps long-term effects on the forest in general. An understanding of the effects on oak regeneration that result from wildlife feeding on acorns or seedlings is very timely because a lack of oak regeneration is a serious issue in some parts of the eastern deciduous forest (Crow 1988).

## Insects and Fungi as Forest Pests

Feeding on foliage by insects in the eastern deciduous forest is not a negative plant-animal interaction under most circumstances. In pre-European times, most foliage-eating insects probably had a negligible impact on forests and instead served as possible thinning agents for the forest vegetation (Spurr and Barnes 1980). Even today, about 5 to 30 percent of the foliage in a forest is lost to feeding by insects in a typical year, with only a minor impact on forest trees (Mattson and Addy 1975). Predators of forest insects, like insectivorous birds, can play a substantial role in offsetting the overall effects of leaf-chewing insects on foliage of trees in the eastern deciduous forest (Marquis and Whelan 1994). Feeding by insects along with other factors, such as disease, however, can combine to replace older, less productive stands with younger, productive stands, thereby "recycling" the forest (Giese 1990).

As discussed in chapter 2, exotic fungi causing chestnut blight and Dutch elm disease nearly eradicated American chestnuts and American elms in the early decades of the twentieth century. However, recent breeding programs have shown promise in restoring American chestnuts and American elms to landscapes of the eastern deciduous forest (Kamalay 1996; The American Chestnut Foundation 1996). But other pests have dramatically affected our eastern forest in more recent times. In 1987, for instance, an estimated 10.6 million hectares of forest trees were killed by insect pests (Council on Environmental Quality 1989), which represented about 3 percent of the total forest acreage in the lower forty-eight states! Two major insect pests in forests of eastern North America are the spruce budworm and the gypsy moth (Cutter, Renwick, and Renwick 1991). The spruce budworm is the principal insect pest in the coniferous forests to the north of the eastern deciduous forest. In 1978, 4.8 million hectares of forest were damaged by this insect. Much needs to be learned in order to effectively reduce the negative impacts of native and exotic pests affecting important tree species in the eastern deciduous forest.

### *Gypsy Moths—An Exotic Pest*

The gypsy moth is a principal, exotic insect pest in the eastern deciduous forest (Cutter, Renwick, and Renwick 1991). This insect has caused extensive defoliation of forests in New England, New Jersey, New York, and Pennsylvania (Figure 4.4). In 1981, for example, about 5.2 million hectares in the eastern deciduous forest were defoliated by this pest. The gypsy moth was introduced into North America from Europe in the latter part of the nineteenth century (Eckess 1982). Trouble with this pest began when Leopold Trouvelot, a French scientist, started a silk industry in New England. He brought the moth to Medford, Massachusetts, in 1869, and it immediately escaped into the wild. By 1910, eastern Massachusetts, Connecticut, Vermont, and parts of Maine and Rhode Island were infested with this pest. Visible defoliation was recorded in

Figure 4.4. Feeding by gypsy moths on leaves of a deciduous tree. (Photograph by E. Alan Cameron, Department of Entomology, Pennsylvania State University.)

New York by 1934 and in Pennsylvania by 1968. Spread of gypsy moths remains uncontrolled today, and infestations now have been discovered in several midwestern, western, and southern states (Skelly et al. 1989). This moth is now in seventeen northeastern states (Sharov, Mayo, and Leonard 1998) and is spreading at a rate of about 20 kilometers per year (Liebhold, Halverson, and Elmes 1992).

Damage to trees by gypsy moths can vary from a reduction in growth rate to death (Powell and Barnard 1982). The severity of damage to trees depends on a variety of factors, such as simultaneous infection by disease (e.g., shoestring root rot, a root-attacking fungus), frequency of defoliation, health of trees before defoliation, topography, tree species, and weather. For instance, after a severe defoliation episode by gypsy moths in a forest stand, 35 percent of the oaks in poor condition had died, whereas only 7 percent of the oaks in good condition were killed (Campbell and Sloan 1977; Houston 1981). After two years of consecutive defoliation in this same stand, 55 percent percent of the poor-condition oaks and 22 percent of the good-condition oaks died within two years. Health of northern red oak stands in Pennsylvania began to decline in the late 1980s immediately after severe defoliation by gypsy moths (McClenahen, Hutnik, and Davis 1997). But this reduction in health of red oaks was preceded by various environmental stresses, such as drought, air pollution, and other insect defoliations, that made individual trees more susceptible to mortality.

Not all tree species in the eastern deciduous forest are equally susceptible to defoliation by gypsy moths; oaks, especially white and chestnut oaks, are preferred species (Skelly et al. 1989). Hence, stands dominated by oaks usually require more potent control methods against gypsy moths than stands composed of other tree species (Appel and Schultz 1994). Moreover, if the biological bacterium insecticide Bt is used to control gypsy moths, chemicals produced by oaks (tannins) act to neutralize the effects of Bt on gypsy moth populations. Chemical changes in oak leaves fed upon by gypsy moth larvae become less palatable to this insect pest in the following year, thereby reducing the future potential impacts of gypsy moth outbreaks in an oak stand (Schultz and Baldwin 1982).

Since the first outbreak of defoliation by gypsy moths in forests of eastern New England, impacts of this pest on oaks have become of particular interest to resource managers in the eastern United States (Houston 1981). Oaks have economic value to the timber industry and, as we saw earlier, are a major provider of acorns for wildlife. In central Pennsylvania, acorn production was markedly lower in years following heavy defoliation of oaks compared to years before defoliation (pre-1981) (Yahner and Smith 1991). A reduction in acorn production resulting from gypsy moth defoliation in Shenandoah National Park affected the ecology of black bears by causing short-term shifts in bear food habits, with bears shifting from acorns to grapes, pokeweed, and other soft mast as food in the autumn (Kasbohm, Vaughan, and Kraus 1995). This gypsy moth infestation and resultant failure of the acorn crop, however, had no detectable short-term effect on bear reproduction or survival in the park (Kasbohm, Vaughan, and Kraus 1996).

Defoliation by gypsy moths can have an indirect effect on birds' nesting success by making nests more vulnerable to detection by predators. For instance, artificial bird nests placed at defoliated sites experienced a significantly higher predation rate (42 percent) than those at nondefoliated sites (23 percent) (Thurber, McClain, and Whitmore 1994). Predation rates on artificial nests on a study area in central Pennsylvania were higher early in the nesting season (38–57 percent) when vegetation was defoliated by gypsy moths compared to later in the season (23–29 percent) when gypsy moths entered the pupal life stage and vegetation had refoliated (Yahner and Mahan 1996a).

Major vertebrate predators of gypsy moths are white-footed mice and several species of forest birds, such as blue jays and yellow-billed and black-billed cuckoos (Smith 1985). In some areas of the gypsy moth range, such as central Pennsylvania, defoliation by gypsy moths is greatest on dry slopes at upper elevations, where abundance and diversity of predators on gypsy moths, including both small mammals and birds, are very low compared to those at lower elevations (Yahner and Smith 1990, 1991). Year-to-year differences in the survival of gypsy moth pupae, in particular, are affected strongly by densities of white-footed mice (Smith 1989). Densities of white-footed mice are associated, in turn, directly with densities of acorns (Elkinton et al. 1996). Thus, management of forest habitat for white-footed mice and other small mammals, by leaving logs, stumps, and cavities (Yahner 1988b), may help mitigate damage by gypsy moths to our forests (Yahner and Smith 1990).

In addition to vertebrate predators of gypsy moths, release of sterile male gypsy moths, chemical pesticides, parasites, invertebrate predators, and pathogens have been used with various degrees of success to control this insect pest (Leonard 1981). In fact, the two principal natural enemies of gypsy moth larvae are fungi and viruses rather than vertebrates (Reardon 1995). A fungus, called "maimaiga" after the Japanese common name for the gypsy moth, was first found in North America in 1989. Since arriving from Japan, maimaiga has spread rapidly throughout much of the range of the gypsy moth in the eastern deciduous forest. In contrast to the gypsy moth fungus, the gypsy moth virus probably was present in populations of gypsy moths brought from Europe to North America by Leopold Trouvelot in 1869 (Onken 1995). Maimaiga is effective against gypsy moth populations regardless of the density of this insect pest, whereas the gypsy moth virus is effective only when gypsy moth populations are high. Larvae killed by the fungi typically hang vertically with head down (Reardon 1995), whereas larvae killed by the virus characteristically hang in an inverted "V" position (Onken 1995).

### *Other Exotic Pests*

In the eastern deciduous forest, four additional exotic pests worth mentioning are beech bark disease, dogwood anthracnose, and two species of woolly adelgids (Williams 1994; McClure, Salom, and Shields 1996; Rabenold et al. 1998). The beech bark disease arrived in North America from Europe and was initially discovered in Nova Scotia in the 1890s (Williams 1994). This disease affects American beech and is caused by the interaction of two types of beech bark fungi and an insect called the beech scale. Feeding holes drilled by the insect in beech trees allow entrance of the fungi; the fungi kill mature trees by girdling. About 88 percent of the mature trees in the Allegheny National Forest of northwestern Pennsylvania, for instance, were killed or negatively affected by this disease by 1994.

A more recent disease, dogwood anthracnose, was first found in New York City and the state of Washington in 1977 (Williams 1994). The fungus causing dogwood anthracnose probably originated in Asia and since has caused extensive mortality of flowering dogwoods throughout the eastern deciduous forest. Dogwoods growing in the moist microenvironment beneath the forest canopy seem to be more susceptible to this disease than those found in more open situations.

The hemlock woolly adelgid (HWA) is an insect native to Japan and perhaps China, but it first appeared in British Columbia during the early 1920s, and in Virginia around the 1950s (McClure, Salom, and Shields 1996). The HWA feeds on needles of the conifer eastern hemlock and often kills the trees within four years. In the eastern deciduous forest, the hemlock is a late successional species that eventually dominates stands in the absence of the HWA. By the late 1990s, however, most hemlock stands have been devastated by the HWA in Shenandoah National Park of Virginia, but hemlock stands have not been as severely affected in the Delaware National Water Gap Recreation Area of New Jersey and Pennsylvania (Mahan et al. 1997). Although eastern hemlock is not a valuable timber species, it is important as wildlife cover and as a landscape tree. Hence, considerable concern has arisen about the potential future loss of hemlock and associated wildlife in the eastern forest, because a similar forest pest, the balsam woolly adelgid, practically has eliminated Fraser fir trees at high elevations in the Great Smoky Mountains (Rabenold et al. 1998). The balsam woolly adelgid was introduced into North America at the beginning of the twentieth century and initially was discovered in the mountains of North Carolina in 1995. Once infected, a Fraser fir dies within two to seven years. Because of fir defoliation by this adelgid, localized populations of several bird species, such as black-capped chickadees, blue-headed vireos, and black-throated green warblers, have declined in the southern Appalachians.

### *Native Insect Pests*

Two native insect pests of trees in the eastern deciduous forest are pear thrips and elm spanworms (U.S. Department of Agriculture 1985;

Brose and McCormick 1992). Since being recognized as a pest in 1980, the pear thrip has become a major concern to maple producers throughout the northeastern United States. It causes severe damage to leaves of sugar maple by reducing the volume and sugar concentration in maple sap (Kolb et al. 1992). In addition to being a source of maple syrup, sugar maple is important as a native tree species to both the hardwood lumber industry and tourist-associated viewing of autumn foliage colors in the northeastern United States (Teulon et al. 1993). No management strategies have been developed for pear thrips, but prescribed fire has been recommended as an effective means of mitigating pear thrip damage by controlling the timing and the successful emergence of adult thrips from forest soils (Brose and McCormick 1992).

The elm spanworm occurs in southern Ontario and the eastern United States, although it is found as far west as Colorado (U.S. Department of Agriculture 1985). Preferred tree species are hickory, oak, red maple, and white ash. Regarded about 100 years ago as a pest of shade trees in larger cities of the East, the elm spanworm now is known to defoliate large expanses of forests. For instance, elm spanworms defoliated nearly 0.5 million hectares of forest in Pennsylvania during 1993, which is about four times the acreage defoliated by gypsy moths in the state that year (Williams 1994).

Some other insect native pests of deciduous trees in the eastern United States are oak leafrollers, forest tent caterpillars, and fall cankerworms (Skelly et al. 1989). The oak leafroller prefers oaks, whereas the forest tent caterpillar and the fall cankerworm attack several species of trees.

Control of insect and fungi pests, both exotic and native, involves a process called integrated pest management (IPM). IPM uses several methods—biological, chemical, and silvicultural—from a variety of disciplines to control insect pests without affecting the environment or human health (Giese 1990). We must be concerned, however, about the impact of any IPM program on other forest biota, particularly the nontarget insects and animals, such as birds that feed on insects (Yahner, Quinn, and Grimm 1985; Hunter 1990). For instance, use of the infamous insecticide DDT during the past decades had detrimental impacts on many types of insects and vertebrates higher in the food chain. The insecticide Bt, which today is in wide use for gypsy moth control, can affect insect species related to the gypsy moth. Thus, care must be exercised with use of any insecticide because it can lead to a reduction in the diversity of forest insect species; some nontarget insects affected by a given insecticide may be beneficial or even crucial as plant pollinators or as natural predators and parasites of noxious forest insects. The impact of these insecticides on forest animals will be discussed further in chapter 9.

## Large Mammals as Forest Pests

### *White-Tailed Deer*

As the deciduous forest in the eastern United States was removed during the nineteenth and early twentieth centuries, the effects of wildlife on forest regeneration became a major worry (Spurr and Barnes 1980). Although many animals feed on woody species in the forest, high populations of one species in particular, the white-tailed deer, have affected forest regeneration in many areas of the eastern United States. Deer populations irrupted following extensive timber removal and enactment and enforcement of hunting regulations in the eastern United States in the early twentieth century (Figure 2.9). Populations of white-tailed deer continue to grow in the contiguous forty-eight states, being absent or rare only in Utah, Nevada, and California (Hesselton and Hesselton 1982); numbers today exceed fifteen million, which is much higher than anytime in the past (Thomas 1990). In addition to suitable habitat, factors contributing to the increase in deer populations in recent decades include reduced number of hunters, more land posted to prohibit hunting, adaptability of deer to nonforested habitat, and lack of natural predators on deer.

High populations of white-tailed deer are having substantial and deleterious ecological effects in many regions of the eastern deciduous forest (Waller and Alverson 1997). For instance, browsing by white-tailed deer on seedlings can eliminate forest regeneration (Figure 4.5). Browse lines devoid of vegetation can be evident from ground level to a height (about 2 meters) reached by deer while feeding. In northwestern Pennsylvania, high deer densities negatively affected the regeneration of understory tree species, such as red maple and northern red oak (Bowles and Campbell 1993). Even if high populations of deer do not eliminate regeneration of tree seedlings, they may delay the time normally required for regeneration (Marquis 1981).

Deer browsing also can change the species composition of the forest. Commercial tree species, such as pin cherry, and ground vegetation, such as raspberry, are selected as food by deer, thereby allowing less-preferred tree species, such as American beech and black cherry, and ground cover, such as grasses and ferns, to dominate forest stands in

Figure 4.5. Browsing by white-tailed deer in a clear-cut stand in north-central Pennsylvania has virtually eliminated all forest regeneration. (Photograph by author.)

northwestern Pennsylvania (Marquis and Grisez 1978; Tilghman 1989). At Cades Cove in the Great Smoky Mountains National Park, deer browsing changed the species composition from one dominated by deciduous species to one favoring conifers (hemlock and white pine) (Bratton 1980). High deer populations in the upper Great Lakes region are having a major impact on many forest species, including eastern hemlock (Mladenoff and Stearns 1993). Many forest managers believe that successful regeneration of forest stands in some areas of the eastern United States will require adequate deer harvests to substantially lower densities, plus measures, such as fenced exclosures, to exclude deer from feeding in regenerating stands (D. deCalesta, U.S. Forest Service, personal communication). The annual timber loss to browsing of hardwoods by white-tailed deer in northwestern Pennsylvania was estimated to be over $56 per hectare (Marquis 1981).

In addition, grazing by white-tailed deer can have a serious effect on herbaceous plants. For example, at Presque Isle State Park in northwestern Pennsylvania, feeding by deer has reduced the annual seed production in a population of an endangered perennial plant, the hairy puccoon, by as much as 90 percent (Campbell 1993). In localized areas, such as Gettysburg National Military Park, densities of deer were about twenty-eight deer per square kilometer in the 1980s, resulting in a dramatic reduction of forest regeneration in historic woodlots and a virtual elimination of farm crop production in the park (Storm et al. 1989; Vecellio, Yahner, and Storm 1994). A conservative estimate of annual damage by white-tailed deer to agricultural crops in the United States is $100 million (Conover 1997).

The impact of white-tailed deer on the eastern deciduous forest is not limited to plants. The abundance of songbirds that nest or forage in understory vegetation may decline in forests heavily browsed by white-tailed deer (Casey and Hein 1983). Densities of songbird species decreased with increasing densities of deer that were maintained in large enclosures in northwestern Pennsylvania (deCalesta 1994). In recently clear-cut stands of north-central Pennsylvania characterized by extensive deer browsing (Figure 4.5), the diversity of songbird species and the relative abundance of snowshoe hares were lower compared to those in stands with less deer browsing (Dessecker and Yahner 1987; Scott and Yahner 1989). Although we are beginning to understand the various consequences of high white-tailed deer populations on regeneration and certain species of wildlife, future research also needs to focus on the indirect effects of deer on nutrient cycling (e.g., nitrogen), net primary production, and other ecosystem processes in the eastern deciduous forest (Hobbs 1996).

Issues surrounding white-tailed deer management are not restricted to extensive forested regions of the eastern United States. Populations of deer have expanded their numbers in suburbia, farmlands, and parks (Flyger, Leedy, and Franklin 1983; Warren 1991). High deer populations in suburbia have become a major problem because of feeding on native and planted vegetation (e.g., vegetable gardens and shrubbery) and greater incidences of deer-vehicle collisions (Decker 1987; Storm et al. 1989; McAninch and Parker 1991) (Figure 4.6).

Deer-vehicular collisions are of rising concern throughout the United States because of property damage to vehicles and incidences of human injuries and fatalities associated with these collisions (Romin and Bissonette 1996). Twenty-six of twenty-nine states that provided roadkill data by year reported increases in deer-vehicular collisions from 1982 to 1991, which averaged 217 percent and ranged from 20 percent (North Dakota) to 2,206 percent (New Jersey). In 1993, a conservative (and probably underrepresented) estimate of deer roadkills was 7.26 million in the United States, causing a total of $1.1 billion in vehicular repair bills or $1,577 per collision in 1993 inflation-adjusted dollars

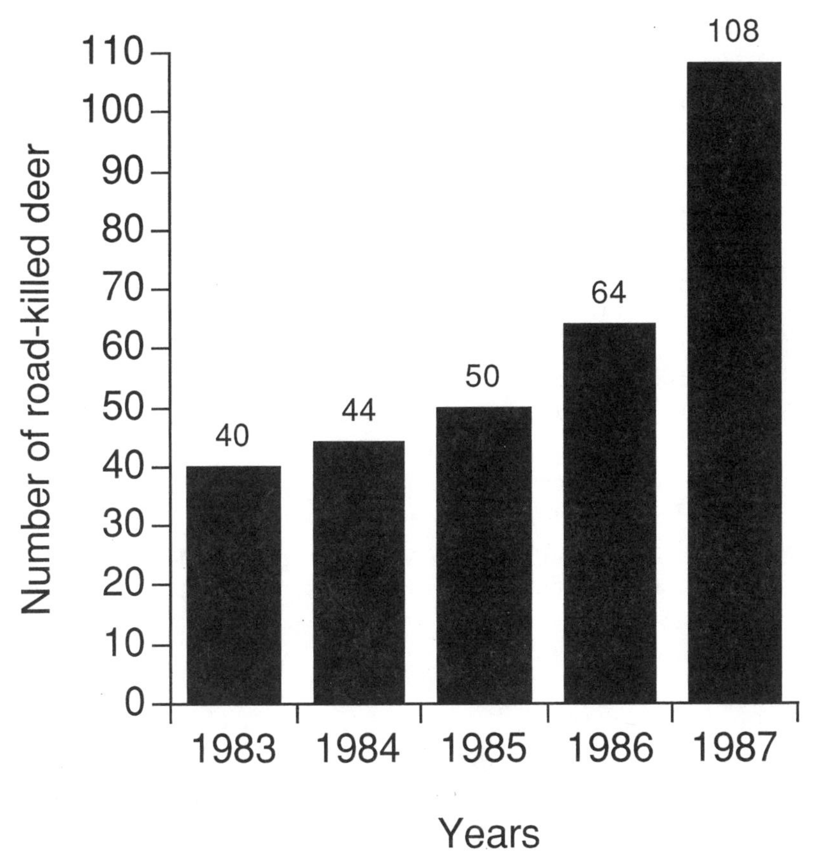


Figure 4.6. Numbers of road-killed deer at Gettysburg National Military Park, Pennsylvania, 1983–87. (Modified from Storm et al. 1989.)

(Conover et al. 1995). From a human perspective, deer-vehicular collisions result in about a 4 percent injury and a 0.03 percent fatality rate nationwide; for deer, colliding with a vehicle is fatal 92 percent of the time (Allen and McCullough 1976; Conover et al. 1995).

Hunting or, in some cases, a controlled reduction program will probably always be the best management tool to control deer populations (Storm et al. 1989; Storm, Cottam, and Yahner 1995). Nontraditional and currently impractical alternative methods of population control, such as deer contraception, are in the early stages of development (Curtis and Richmond 1992). Contraception as a population control in some other species of wildlife, however, has been studied for a couple of decades (Kirkpatrick and Turner 1985); for example, long-term infertility using safe contraceptive methods has been achieved in wild horses (Kirkpatrick et al. 1992). When successful methods of contraception in white-tailed deer are developed, they eventually may have future management applications to populations in parks or preserves where hunting is prohibited or illegal (Kirkpatrick and Turner 1995; Turner, Kirkpatrick, and Liu 1996).

As deer populations increase near centers of high human populations and in outdoor recreational areas (e.g., state forests and parks), the issue of Lyme disease must be considered (Decker 1987). Lyme disease and other human-wildlife interactions will be discussed later in this chapter. Thus, for a variety of reasons, the management of the white-tailed deer, which is perhaps the most popular and aesthetic wildlife species, poses unique problems and issues for natural resource managers in the eastern deciduous forest (Warren 1991; Curtis and Richmond 1992; Diefenbach, Palmer, and Shope 1997). Moreover, this species is certainly the most valuable wildlife species from an economic perspective; in 1991 alone, $7 billion was spent on deer-related recreation in the United States (Conover 1997)!

### *Other Large Mammals*

The white-tailed deer is not the only large mammal affecting plant and animal communities in certain areas of the eastern deciduous forest. The sika deer, an exotic species about two-thirds the size of our native white-tailed deer, has become relatively common in Maryland and Virginia since its introduction in 1916 (Feldhammer 1982). Ironically, the sika deer is an endangered species and is considered a "national monument" in Japan, yet it has become a pest in the United States and elsewhere in the world. This deer may even be displacing the native white-tail in certain areas of Maryland and Virginia, because the sika deer has more generalized food habits and habitat use. Like the white-tail, the sika deer damages forest regeneration by feeding on seedlings and can cause extensive destruction of farm crops.

Domestic cows and wild pigs also can have a negative impact on the eastern deciduous forest. Free-roaming cattle in woodlots browse on woody vegetation and graze on herbaceous vegetation, thereby potentially reducing the diversity of plant life and in turn influencing the associated animal life (Bratton 1980; Casey and Hein 1983). The wild, or feral, pig (hog) is an exotic species from Europe that escaped from barnyards and shooting preserves in the United States (Wood and Barrett

1979; Howe, Singer, and Ackerman 1981; Sweeney and Sweeney 1982). Populations of feral pigs now occur in forests of the southeastern United States and have become well established since the 1930s in some areas, such as the Great Smoky Mountains National Park. Pigs cause serious damage to natural plant communities by their feeding on gray beech, tulip poplar, and spring beauty; furthermore, pigs are potential competitors with native wildlife for acorns, and they feed on eggs and young of ground-nesting birds, such as eastern wild turkeys.

## Human-Wildlife Diseases

### *Lyme Disease*

Lyme disease, rabies, and hantavirus are three important diseases transmitted from wildlife to humans. Lyme disease is caused by a tick-borne bacterium that initially was diagnosed near Lyme, Connecticut, in 1975 (Steere et al. 1983; Ostfeld 1997). This disease has become one of the most serious public health concerns in the United States, especially along the East coast and in the upper Midwest (Booth 1991; Barbour and Fish 1993). It has now been reported in forty-eight states (Ostfeld 1997). The numbers of cases of Lyme disease have risen dramatically since the early 1990s (Ginsberg 1994); for instance, more than eighteen thousand cases were reported in 1996, which was an increase of 41 percent from 1995 (Fraser et al. 1997).

Lyme disease was described in Europe in the early twentieth century, and it perhaps existed in both Europe and North America for thousands of years (Ostfeld 1997). The disease probably went unnoticed because of the prevalence of more serious diseases historically and because health-care workers in more recent decades were unfamiliar with its symptoms. Often a person develops a circular rash at the site of the tick bite (Ostfeld 1997). Early symptoms of the disease in humans include muscle aches, low-grade fever, and lethargy; these are followed later by arthritis and damage to the heart and nervous system (Barbour and Fish 1993; Gage, Ostfeld, and Olson 1995). Lyme disease has minimal or no effect on wildlife but may cause fever, arthritis, and lameness in domestic animals, such as dogs, cats, and cows (Ginsberg 1994).

The white-tailed deer and, in particular, the white-footed mouse are two major hosts of the deer tick, which transmits the bacterium causing Lyme disease in humans and other mammals (Figure 4.7). Larval deer ticks rarely are infected with the Lyme disease bacterium, with greater

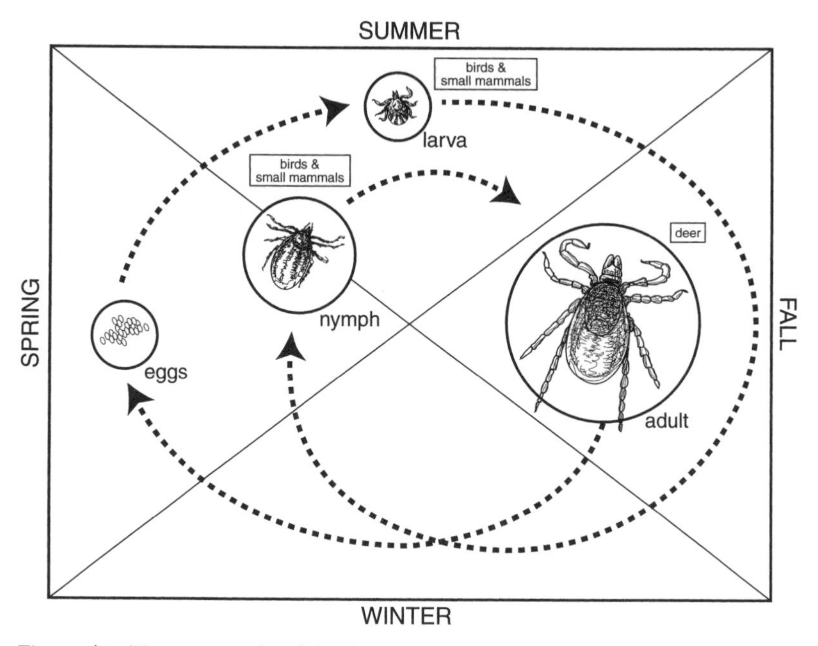


Figure 4.7. Two-year cycle of the deer tick, showing major hosts for each of the three life stages. (Modified from Van Buskirk and Ostfeld 1996.)

than 99 percent of the larval ticks hatching from eggs free of the disease (Gage, Ostfeld, and Olson 1995; Ostfeld 1997). Larval ticks acquire the bacterium by feeding on infected hosts, such as the white-footed mouse, in August and September after the mice became infected earlier by nymph ticks in June and July. In general, adult deer ticks usually feed on larger hosts, like deer, whereas juvenile (nymph and larva) ticks usually feed on various smaller mammals and birds (Van Buskirk and Ostfeld 1995). Although each of these three life stages of deer ticks can parasitize humans, most cases of Lyme disease in humans probably are transmitted by nymphs, because they are smaller than adults, feed for a shorter time period, and are thus harder to detect than adult ticks; furthermore, humans are more active in summer when nymphs are more active than in autumn and spring when adult ticks search for hosts (Gage, Ostfeld, and Olson 1995). In areas with Lyme disease epidemics, 50 to 70 percent of the adult ticks and 25 to 35 percent of the nymphs may be infected with the bacterium (Ostfeld 1997).

Increases in white-tailed deer populations and expansions in the range of deer tick populations during the past few decades possibly have contributed to higher incidences of Lyme disease in humans (Ginsberg

1994). Although deer seemingly may be the most feasible species to manage in order to mitigate the spread of the disease because they are hunted, a lowering of deer populations has not been shown to reduce incidences of Lyme disease (Wilson and Deblinger 1993; Van Buskirk and Ostfeld 1995). Habitat manipulation by mowing or burning may temporarily lower tick populations, but tick densities can increase within a period ranging from months to a couple of years (Wilson 1986). Instead, the control of Lyme disease may be dictated by the availability and the density of competent hosts, especially the white-footed mouse (Van Buskirk and Ostfeld 1995; Ostfeld 1997). Therefore, the transmission of the Lyme disease bacterium to other wildlife species and humans conceivably may be slowed by increasing the diversity and the abundance of other small mammals, such as shrews and chipmunks, which are less competent hosts, i.e., less likely to harbor the bacterium (Fish and Daniels 1990). Therefore, if the small mammal community consists of a variety of species, infected ticks then would feed on both competent and less competent hosts, thereby diluting the spread of the disease in the wild (Van Buskirk and Ostfeld 1995). One way of increasing the diversity and the abundance of small mammals for the control of Lyme disease may be to provide habitat diversity at local and regional levels (e.g., Anthony, Niles, and Spring 1981; Yahner 1988b).

A better understanding of plant-animal interactions in the eastern deciduous forest may increase our ability to predict year-to-year variations in occurrence and potential impacts of Lyme disease on humans. For example, one possible scenario recently proposed is based on the relationship among deer, mice, and acorns, which goes as follows (Ostfeld, Miller, and Hazler 1996; Ostfeld 1997): The abundance of larval deer tick populations has the potential to increase in oak forests following years of good mast production, because high acorn crops in autumn can attract deer to these feeding areas (e.g., McShea and Schwede 1993). Some of these deer are likely to carry adult ticks infected with the Lyme disease bacterium, and larvae produced by these adult ticks then could infect resident white-footed mice. Because the deer-tick life cycle is two years in length (Figure 4.7), the risk of Lyme disease to humans in oak forests would reach a peak two summers following the high autumn acorn crop.

Lyme disease can largely be prevented in humans by personal protection against tick bites (Gage, Oltfeld, and Olson 1995; Ostfeld 1997). When entering areas infested with deer ticks, people should cover exposed areas, e.g., wear long-sleeved shirts and tuck pant legs into socks, apply insect repellents containing DEET, and wear light-colored clothing. During and after entering these areas, the body should be inspected carefully for ticks. Lyme disease can be treated with antibiotics, and a vaccine for humans (LYMErix) has been approved by the Food and Drug Administration as of January 1999.

### *Rabies*

Rabies is a disease known to humans since at least 500 B.C. (Krebs, Wilson, and Childs 1995). The Latin word *rabies* is believed to mean "to do violence" because of aggressiveness shown by some animals infected with the virus. The rabies virus often is transmitted via saliva into cuts or wounds in the skin caused by bites (Centers for Disease Control and Prevention 1991); the virus very rarely has been transmitted airborne in bat-infested caves (Constantine 1967). The rabies virus infects the nervous system and initially causes pain at the wound site, fever, and headaches in humans (Fishbein 1991). Later symptoms of rabies in humans include disorientation, hallucination, agitation, and, rarely, aggression; the disease is nearly always fatal.

Principal rabies hosts in the eastern deciduous forest of North America are red and gray foxes, raccoons, striped skunks, and some bats, e.g., the big brown bat (McLean 1970; Smith 1989; Krebs, Wilson, and Childs 1995). Rabies is extremely rare in rabbits and rodents, with the exception of woodchucks. The disease is much more prevalent in wild than in domestic animals in the United States and in other developed countries, whereas the domestic dog is often the major reservoir for the rabies virus in undeveloped countries. In the United States, cats are the most common domestic species reported with rabies.

Rabies initially was recorded in North America in Virginia in 1753 (Carey 1982). The relative occurrence of rabies in wild animals in the United States has changed over the past few decades (Krebs, Wilson, and Childs 1995). Foxes were more likely to be rabid than other types of wildlife prior to the 1960s, whereas rabies in skunks predominated in the next two decades. Since the early 1990s, however, rabies is most prevalent in raccoons, especially in northeastern and southeastern states.

Rabies has been best studied in the red fox (Preston 1973; Carey 1982). About 40 percent of rabid red foxes show a furious form of behavior, which includes aggressive attacks and biting; 60 percent of rabid foxes manifest a dumb form, whereby animals are visibly docile and weak, and lack control of jaw musculature. Outbreaks of rabies in foxes occur in fall and winter, corresponding to dispersal of young from family groups and the breeding season, respectively. The symptomatic period averages about seven days, and foxes exhibiting the furious form can travel as much as 30 to 50 kilometers. While wandering during the symptomatic period, rabid foxes are likely to encounter other foxes because foxes' home ranges are relatively small. Thus, red foxes are an effective vector of rabies compared to gray wolves, which have very large home ranges. If a member of a wolf pack becomes infected, the entire pack becomes decimated rather than spreading the virus to other packs (Chapman 1978).

In theory, disease can potentially regulate population sizes in animals. However, no published literature supports the fact that rabies acts to control populations of skunks, foxes, or other mammals (Schubert et al. 1998). Oral vaccination using baits has been successful in reducing the incidences of rabies in red fox and raccoon populations (Wandeler 1991; Rupprecht et al. 1993; Muller 1994; Tinline, MacInnes, and Smith 1994), but intramuscular vaccination did not decrease rabies in skunk populations (Schubert et al. 1998).

Today, the occurrence of rabies in humans in the United States is much less frequent than in the past. In the early 1900s, incidences of rabies in humans exceeded 100 per year but have declined to less than one per year since the 1960s (Krebs, Wilson, and Childs 1995). Rates of rabies transmission from wild animals to either domestic animals or humans in recent decades have been reduced because of vaccination of pet animals. Rabies can be treated effectively after humans have been exposed to the virus, and all bites by foxes, skunks, raccoons, and bats should be considered possible exposures requiring treatment (Krebs, Wilson, and Childs 1995). Preexposure vaccination is recommended for persons, such as mammalogists, who may be in contact with wildlife species, e.g., foxes, that typically transmit the disease.

### *Hantavirus*

Another disease transmitted from wildlife to humans is hantavirus pulmonary syndrome or, simply, hantavirus (Childs, Mills, and Glass 1995). Hantaviruses first came to the attention of physicians in the United States during the Korean War when United Nations troops were diagnosed with the virus. (The name of the viruses originated from the Han-Gang River of Korea.) The first outbreak of hantavirus in North America was reported in the Southwest in 1993 (Zeitz et al. 1995); as of

April 1998, 179 deaths have been attributed to the disease in the United States, and isolated reports of hantavirus have continued to occur in recent years (The Associated Press Release, 13 April 1998). In marked contrast, hantaviruses annually cause several hundred thousand human deaths each year in Europe and Asia (Childs, Mills, and Glass 1995).

Hantavirus in North America results in respiratory rather than kidney failure, with the latter characteristic of cases in Europe and Asia (Duchin et al. 1994). Early symptoms of hantavirus in North America include fever, vomiting, nausea, coughing, and difficulty in breathing; in a few hours or days, respiratory disease occurs; more than 50 percent of the people afflicted with hantavirus in North America have died. Hantavirus likely is transmitted from rodents to humans when victims inhale small particles of dust laden with rodent urine, feces, or saliva (LeDuc 1987). Hantaviruses have been found in several rodent species, such as deer mice, hispid cotton rats, Norway rats, and meadow voles. Outbreaks of hantavirus seem to be associated most often with rodents in rural settings, as with the initial outbreak observed in Navajos of New Mexico in spring 1993 (Zeitz et al. 1995).

Incidences of hantavirus may be reduced by making habitat less suitable for rodents near human dwellings, e.g., removing brush piles near buildings (Childs, Mills, and Glass 1995). Attempts should be made to make human dwellings inaccessible as homesites for rodents, food in human dwellings should be made unavailable to rodents, and population control of rodents in dwellings may be necessary. People trapping or handling rodents potentially infected with hantavirus should take special precautions, such as using respirators, disinfectants, and thick rubber gloves (Mills et al. 1995).

In summary, an understanding of the interactions of wildlife with vegetation and humans in the eastern deciduous forest is important for a better appreciation of the ecology and the conservation of forest resources. Some of these interactions, like those created by high populations of gypsy moths and white-tailed deer, present some of the most challenging problems facing natural resource managers concerned about succession and management in the eastern deciduous forest today and in the future. Forest succession and management will be discussed in the next chapter.

# 5. FOREST SUCCESSION AND MANAGEMENT

## Forest Succession

Forest succession is the regrowth of a forest stand unaided by artificial seeding, planting of seedlings, or other human activities following a natural or human-induced disturbance (Cutter, Renwick, and Renwick 1991). During succession, the species composition and the abundance of plant and animal species change over time, often in a somewhat predictable fashion, as habitat conditions become more suitable for some species while less so for others (Aber 1990; Barnes 1991) (Figure 5.1). Succession, an important ecological concept, must be understood to explain patterns of the distribution and the abundance of plants and animals in the eastern deciduous forest.

### *Plant Succession*

Shade-intolerant or pioneer species often predominate in early-successional stages soon after a mature forest stand is disturbed. These species are called pioneer species because they are the first to germinate and grow rapidly (Spurr and Barnes 1980). Examples of pioneer plants in the eastern deciduous forest are jack pine, aspen, grasses, and forbs (Hocker 1979; Aber 1990; Patton 1992). Pioneer plant species are replaced after a time by shade-tolerant, more competitive, and longer-lived plants. Shade-tolerant plants invade (via wind or animal dispersal) the disturbed area or were present but growing at a slower rate than the pioneer plants (Spurr and Barnes 1980; Aber 1990). Seedlings of shade-tolerant trees

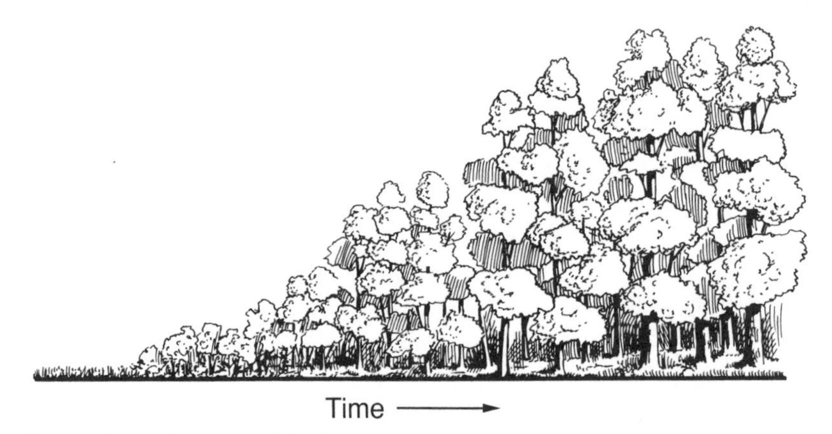


Figure 5.1. Schematic of the forest succession of plants following a natural disturbance such as a wildfire.

become established beneath the forest canopy and eventually penetrate the canopy as overstory trees; examples of these types of trees in the eastern deciduous forest are sugar maple, American beech, and eastern hemlock. Other shade-intolerant species, more appropriately termed gap-phase species, germinate under the forest canopy but only grow as overstory trees after an opening (i.e., a gap) is created in the canopy, such as by the death of an overstory tree, thereby allowing sunlight to reach the forest floor. Gap-phase species in the eastern deciduous forest are represented by black cherry, white oak, northern red oak, and white pine.

As a forest stand undergoes changes in the dominance of tree species during succession, it typically proceeds through five general structural or growth phases: seedling, sapling, pole, mature or sawtimber, and old growth (Patton 1992). In the eastern deciduous forest, seedlings of shade-intolerant species, such as aspen, tulip poplar, and black cherry, are the first to become established and grow to the mature phase (Aber 1990; Gilliam, Turrill, and Adams 1995). Over time, the structural phases of these species are replaced with those of species intermediate in shade tolerance, for example, white oak, northern red oak, and yellow birch. Finally, the forest stand becomes dominated by mature tree species that are very shade-tolerant, like red maple, sugar maple, and American beech.

Succession may be classified as either primary or secondary (Spurr and Barnes 1980; Aber 1990). Primary succession occurs immediately after a catastrophic disturbance that leaves a site virtually devoid of soil and both plant and animal life, as with the volcanic eruption of Mount

St. Helens in 1980 (Frenzen 1992). In the blowdown zone of Mount St. Helens where all old-growth trees were leveled by the force of the blast, only three of twenty species of small mammals survived to repopulate the area (Andersen and MacMahon 1985). Secondary succession follows a less severe disturbance of the forest, such as a fire or timber harvest, that disrupts rather than completely destroys the biotic community. Primary succession may take hundreds or even thousands of years, whereas secondary succession may take only decades or a few hundred years (Aber 1990). A distinction between primary and secondary succession, however, is arbitrary because succession is an ongoing process in any forest. Even without a disturbance, a forest undergoes continual succession because of subtle changes in microclimate or other environmental factors. Furthermore, insect defoliation, windthrows, and other events occur periodically in a forest. These events constantly change the structure and sometimes even the species composition of a forest with death of individual trees (Spurr and Barnes 1980; Patton 1992).

### *Animal Succession*

As the plant community of a forest stand changes via succession, the animal community also varies in response to changing habitat conditions. In early- or midsuccessional forested habitats, wildlife species, such as common yellowthroats, eastern cottontails, and white-tailed deer, typically occur. As plant succession proceeds further, these species are eventually replaced with other wildlife that are characteristic of mature forests, such as scarlet tanagers, southern flying squirrels, and black bears (Figure 5.2).

Natural resource managers must understand changes in the distribution and the abundance of forest plant and animal species resulting from succession to ensure suitable habitat for a diversity of forest species. Although the variety of plant and animal species generally is greater in older forest stands than in younger stands (Yahner 1986a; Gullion 1990), maximum diversity of biota in a region may be achieved only by providing an assortment of forest stands of different ages (Hunter 1990). Moreover, a thorough appreciation of patterns of forest succession will better enable natural resource managers to predict and possibly mitigate the impacts of natural and human-induced disturbances on forest biodiversity. Two major types of disturbances influencing secondary succession in the eastern deciduous forest are fire and timber harvesting (Spurr and Barnes 1980).

## Fire and Succession

### *Prevalence of Fires*

Wildfires have been a natural force affecting the eastern deciduous forest since long before humans set foot on North America. Before European settlement, fires were a relatively frequent event in the eastern forest. For example, in northern Minnesota before the seventeenth century, fires probably occurred every fifty to one hundred years (Lorimer 1990a). In southwestern New Hampshire, fires historically were less common in the landscape, occurring at intervals of 150 to three hundred years (Henry and Swan 1974; Oliver and Larson 1996). After European settlement but before the twentieth century, fires were used extensively in the eastern deciduous forest as a means of clearing land for agriculture and homesites (Lorimer 1990b). During most of the twentieth century, numbers and sizes of forest fires have been reduced throughout North America.

Natural fires in today's eastern deciduous forest occur about every 100 to 400 years (Cutter, Renwick, and Renwick 1991). Many present-

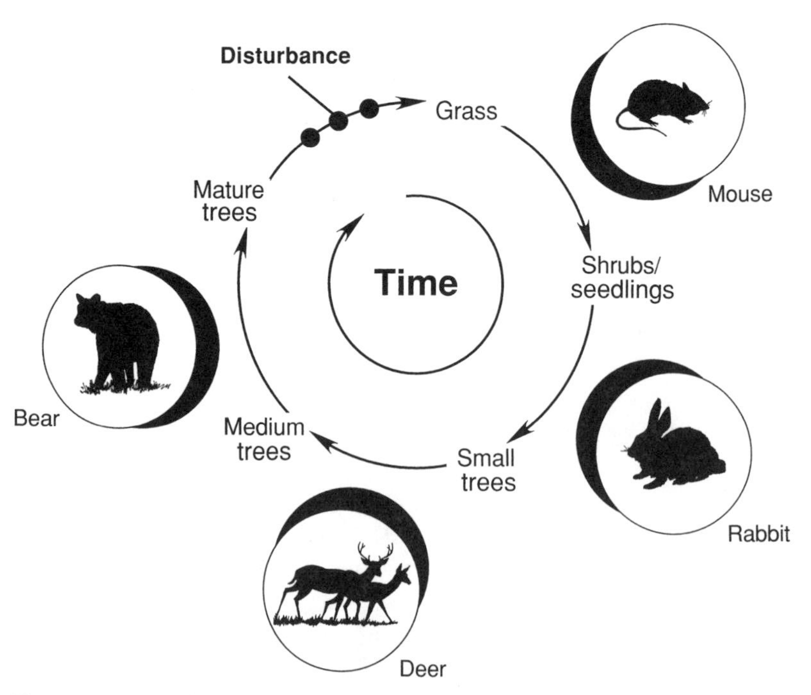


Figure 5.2. Animal succession proceeding from an early-successional (recently cut) forest stand to a mature forest stand.

day wildfires (26 percent) in the United States are set by laypersons to produce cattle forage, to kill chiggers, ticks, or snakes, or to produce wildlife or hunting habitat, particularly in the southern states. Other fires are caused by careless smoking habits (19 percent), debris burning (18 percent), lightning (9 percent), machine use (8 percent), and campfires (6 percent) (Lorimer 1990b). More than six thousand natural fires were caused by lightning in the national forests of the United States alone between 1960 and 1969, but most of these were suppressed by firefighters and were restricted to 4 hectares or less. The reality of extensive fires set by lightning or by people and their dramatic effects on forests, wildlife, and people are occasionally brought vividly to our attention by the media, as was the case with fires in Yellowstone National Park and southern California in 1988 and 1993, respectively (Beebe and Omi 1993).

Forest fires can be divided broadly into two groups: crown and surface fires (Oliver and Larson 1996). Crown fires burn large trees, whereas surface fires burn vegetation and woody debris on the forest floor. Dry weather conditions and high concentrations of fuel (e.g., dead and dying woody material) predispose a forest stand to both crown and surface fires.

### *Effects of Fires on Plants*

Fires are beneficial to the germination of certain deciduous and coniferous tree species in the eastern forest. Tree species promoted by fire typically have developed adaptations, such as thick, fire-resistant bark or seeds that germinate only after being exposed to intense heat from fires (Lorimer 1990b). Trees without a layer of thick, dead outer bark and most other plants are very susceptible to heat from forest fires. Temperatures of 64°C are instantly lethal to all plant tissue (Van Lear and Waldrop 1989).

Aspens and oaks are examples of deciduous tree species in the eastern deciduous forest that are benefited by fire. The root suckers of aspen clones often do not germinate successfully without fire (Schier 1975) (Figure 5.3). Most oak species survive and grow well on upland forested sites exposed to periodic fires throughout the eastern states. Mature oaks have thick bark, and seedbeds created by fires on the forest floor favor acorn germination (Lorimer 1985). Younger oaks in the forest understory have a remarkable ability to resprout after a fire compared to other tree species, in part because fires help control competing hardwood tree

Figure 5.3. Aspen clone in Pennsylvania. (Photograph by author.)

species (Van Lear and Waldrop 1989; Van Lear 1991). Fires also can delay the succession of oak forests to maple forests by causing high mortality of maples, which have a relatively lower resistance to fires than oaks; maples usually are late-successional species in the eastern deciduous forest (Abrams 1992). The lack of fire may possibly contribute to future declines in oaks in some areas of the eastern deciduous forest, as in Pennsylvania (Abrams and Ruffner 1995).

Fires have enabled white, jack, red, and pitch pine to thrive in forests of the northeastern United States and southeastern Canada. Similarly, loblolly, shortleaf, jack, and slash pine are well adapted to conditions created by fires in forests of the southeastern United States. Many of these pines have tightly closed cones, termed serotinous, that open and release seeds only after high temperatures created by a wildfire melt the resin sealing the cones (Lorimer 1990b). Prescribed fires are sometimes deliberately set by foresters to prepare seedbeds for fire-resistant pines, e.g., loblolly pine, by killing other plant life that might compete with the maintenance of pine plantations (Cutter, Renwick, and Renwick 1991).

### *Effects of Fires on Animals*

Unlike plants, animals do not have morphological or reproductive adaptations to forest fires. Animals readily suffocate during a wildfire when temperatures reach or exceed 63°C and oxygen supplies become depleted (Howard, Fenner, and Childs 1959). Smaller and less mobile animals, such as mice on or near the forest floor, can succumb directly from conditions created by intense fires (Bendell 1974). Even if small animals escape death from a recent wildfire, they may be forced in the short term to forage in areas with little or no vegetation, making them more conspicuous to predators, like hawks and owls (Patton 1992). Interestingly, the agouti coloration of fox squirrels in the southeastern United States perhaps evolved as a means of concealing squirrels from predators foraging on the forest floor; the coloration of these squirrels closely resembles the color of charred stumps and logs following a wildfire in southern pine forests (Kiltie 1989).

Some animals modify their behavior to escape the danger of wildfires. Burrowing small mammals, such as chipmunks, may seek refuge in underground sites. Although the effects of fire on amphibians and reptiles are not well known (Patton 1992), amphibians potentially can escape lethal fire conditions by remaining in mud and water (Komarek

1969). Death by fire of birds and large mammals is probably rare because these species are relatively more mobile than other wildlife (Patton 1992). Fires, however, can destroy nests of birds and eliminate insects and other foods important to birds and mammals.

### *Indirect Effects of Fires*

High-intensity fires can have negative indirect effects on the eastern deciduous forest ecosystem by augmenting soil erosion and runoff of rainfall on burned areas (Lorimer 1990b). These conditions can cause sediment problems in local streams and lakes to the detriment of aquatic plant and animal life. Moreover, removal of vegetation near streams by fire can increase the water temperature, which in turn can lead to changes in species composition and distribution of fish and other aquatic life (Patton 1992).

Low-intensity fires, on the other hand, provide indirect benefits to the eastern deciduous forest ecosystem by releasing and cycling nutrients stored in dead timber, leaf litter, and other plant material, which then become available for plant uptake (Lorimer 1990b; DellaSalla et al. 1995) (Figure 5.4). A low-intensity fire can reduce the biomass of organic mate-

Figure 5.4. A low-intensity fire in a northern pin oak stand in central Wisconsin. (Photograph by Marc D. Abrams, School of Forest Resources, Pennsylvania State University.)

rial on the forest floor by as much as 64 percent, which can consist of 5,600 to 6,720 kilograms per hectare of material (Moehring, Grano, and Bassett 1966; Brender and Cooper 1968). When dead plant material is burned periodically by a low-intensity fire rather than allowed to accumulate in the forest, the threat of a severe wildfire is diminished (Lorimer 1990b; Cutter, Renwick, and Renwick 1991). Furthermore, removal of dead plant material can reduce the susceptibility of a forest to damage by wind, diseases, and insects (Spurr and Barnes 1980).

Fires can modify the structure of a forest stand by reducing its canopy cover. The resulting increase in light in turn facilitates the growth of pioneer species of herbaceous plants, shrubs, and trees on the forest floor. A layer of lush, ground-level vegetation provides suitable habitat conditions for a variety of wildlife, such as white-tailed deer, snowshoe hares, and black bear (Bendell 1974).

Prescribed fires, which are deliberately set and controlled, are used occasionally by foresters to remove woody debris on the forest floor following a timber harvest or to prepare a seedbed for planting of tree species that require bare soil (Lorimer 1990b). In addition, wildlife managers in northern states use prescribed fires to create favorable habitat for ruffed grouse, which need a mixture of forest stands of different ages to meet their seasonal food and cover requirements (Gullion 1972). Similarly, prescribed burning of pine stands at one- to two-year intervals has been used in southern states to produce habitat for northern bobwhite (Stoddard 1931; Van Lear and Waldrop 1989). The fires stimulate the growth of plants that are important as food to bobwhites and increase the biomass of insects available to bobwhite broods. Burning as a habitat management tool may help to reverse population declines of bobwhites in the South (Brennan 1991). The extinct heath hen, related to ruffed grouse and northern bobwhite, probably also relied on habitat created by wildfires, but exclusion of fires by humans in the Northeast during the early twentieth century likely contributed significantly to the demise of this bird (Thompson and Smith 1970; Matthiessen 1978). The availability of woody browse as food for white-tailed deer and elk can be increased in some forest cover types, such as oak–shortleaf pine, by using prescribed fires at three-year burn intervals (Masters 1991). Furthermore, prescribed fires may be valuable in maintaining plant diversity in a rare and unusual ecosystem, the serpentine barrens, which occurs on serpentine bedrock in eastern North America (Arabas 1997).

*Forest-Fire Policies*

Although fires apparently played an integral role in the ecology of the eastern deciduous forest before the twentieth century, their importance to the forest of today has been overlooked. The Smokey Bear educational campaign by the U.S. Forest Service and the National Advertising Council has been very successful in limiting the number and the extent of forest fires in the United States. Yet most natural resource managers now agree that fire is an essential part of a fire-adapted forested ecosystem (Lorimer 1990b; Miller 1990). Large fires have been proposed as a means of restoring natural disturbances in wildland landscapes subjected to fire suppression for many decades (Baker 1994). But based on simulation models, restoration of wildlands using a natural fire regime may require fifty to seventy-five years. Fires carelessly set by the public, however, must continue to be prevented. In the northeastern states, careless smoking is the leading cause of wildfires, which often become large and uncontrollable and may lead to irreparable damage to the forest (Lorimer 1990b). The challenge then is to convince the public of the importance of managed fires to the ecology of the forest while effectively informing them about wildfire prevention (Manfredo et al. 1990).

From about the 1920s until 1972, the policy of federal agencies, such as the U.S. Forest Service and the National Park Service, was to fight all natural fires (Cutter, Renwick, and Renwick 1991). Fires that burned understory vegetation and organic material on or near ground level are usually easy to fight, particularly during wet or low-wind conditions. Conversely, fires started during dry and windy conditions are difficult to extinguish, especially where the buildup of dead timber and other plant material is abundant. Generally, these fires become high-intensity crown fires that burn in the treetops and kill virtually all vegetation in the forest stand.

In 1972, federal agencies initiated a "let-burn" policy for wildfires (Cutter, Renwick, and Renwick 1991); this let-burn policy was tested to the limit during an extremely dry period in 1988 when about seventy-two thousand fires burned more than 1.6 million hectares in twenty-three states. Nearly 25 percent of this acreage was in Yellowstone National Park, where approximately 45 percent of the park was burned (Lorimer 1990b). Despite the apparent devastation caused by the Yellowstone fires, only 0.1 percent of the park was destroyed to the point that it will never fully recover to the prefire conditions. Today, the park is green, and wildlife habitat is abundant and improved (Miller 1992). Because of the extensive Yellowstone fires, the National Park Service determined that its let-burn policy had to be modified when a fire becomes extensive and dangerously uncontrollable (Elfring 1989).

In September 1988, the U.S. Forest Service and the National Park Service established the Fire Management Policy Review Team to review national fire policies in wilderness and national parks and to recommend actions to mitigate problems experienced in the 1988 dry season (U.S. Department of the Interior 1989). Later in 1995, five federal agencies, including the Forest Service, Bureau of Land Management, National Park Service, Fish and Wildlife Service, and Bureau of Indian Affairs, developed the Federal Wildland Fire Management Policy and Program Review Report (U.S. Department of the Interior 1995). This policy report has resulted from recognition that the proper use of wildfires is an important tool for the ecology and management of ecosystems. Two major goals of the policy report are to use wildfires for restoring and maintaining healthy ecosystems, while minimizing undesirable effects and educating the public and agency personnel about the role of fire in the natural functioning of ecosystems. In addition, this report established goals and actions to ensure that federal agencies work in conjunction with state and local governments, thereby minimizing fire hazards and risks to urban areas. In 1996, various coordinating groups comprising personnel from federal and state agencies met to begin a discussion of how to implement goals and actions given in the 1995 policy report (U.S. Department of the Interior 1996a).

Forest fires, in general, are responsible for about 8 percent of all atmospheric pollution in the United States (see chapter 9). Wildfire emissions are regulated by federal air quality programs under the Clean Air Act, and the U.S. Environmental Protection Agency is mandated to establish air quality standards favorable to human health (U.S. Department of the Interior 1995). Furthermore, use of wildfires can potentially increase levels of atmospheric carbon dioxide and, therefore, is a concern from a global climatic change perspective. Although low-intensity fires may have little effect on soil erosion and water quality (Brender and Cooper 1968), the impact of any fire on air quality in the vicinity of a dense human population center is an issue (Lorimer 1990b). Hence, the U.S. Forest Service has developed smoke-management guidelines to minimize the impacts of prescribed fires on atmospheric quality (U.S. Department of Agriculture 1976). Because of these guidelines and legitimate environmental concerns about the use of prescribed fires, the use of such fires as a management tool is limited in the eastern deciduous forest (Van Lear and Waldrop 1989).

## Timber Harvesting and Succession

Timber harvesting is an important factor affecting secondary succession in the eastern deciduous forest. Forest management involves both silviculture, which is the planting, growing, and tending of a stand of trees for later harvest, and the economics of growing trees (Society of American Foresters 1981). The term *silviculture* is derived from the word *silva,* which means "forest" in Latin. Today, sound methods of forest management in the eastern deciduous forest should be used not only to harvest timber but also to manage for wildlife, watersheds, and recreational opportunities (Patton 1992).

### *Forest-Management Practices*

Forests are managed for timber production using two basic systems: uneven-aged management and even-aged management (Society of American Foresters 1981; Lorimer 1990a). Uneven-aged management is a system in which trees of different ages are maintained by a harvesting practice called selective cutting. Selective cutting creates frequent (usually every five to ten years), small-scale disturbances in the forest stand. Because uneven-aged management maintains trees of many sizes and ages, it better approximates the process of natural regeneration in a forest than does even-aged management.

Even-aged management, in contrast, is a system in which trees of a desired age are harvested at the same time and then are allowed to regenerate naturally or artificially. This practice results in trees of approximately the same age and size, usually within a relatively large area. Even-aged management produces less frequent, larger-scale modification of the forest stand compared to uneven-aged management (Hunter 1990; Lorimer 1990a). Time between harvests of trees is termed a rotation. A rotation can vary from as little as twenty to thirty years for timber grown for pulp or plywood to at least seventy-five years to produce high-quality lumber. The overall goal of even-aged management has traditionally been to produce timber economically and within the shortest time period (Society of American Foresters 1981). Three traditional methods of timber harvest using an even-aged management system have been employed in the eastern deciduous forest: clear-cutting, shelterwood, and seed-tree methods. In addition to these harvest methods, some alternative even-aged management systems have been developed in recent years, such as the two-age cutting method.

### *Selective-Cutting Method: Single-Tree and Group-Tree Selection*

The selective-cutting method is the removal of single trees or a small group of trees from a stand of intermediate-aged or mature trees. These methods are termed single-tree and group-tree selection, respectively (Society of American Foresters 1981) (Figure 5.5). Both types of selective cutting allow seedlings and sprouts to germinate and grow in the space provided by the tree removal. Because selective cutting removes relatively few trees at a given time, species that germinate and grow in the recently established forest opening are those that tolerate low-light conditions. Single-tree selection in the northern hardwood forest region benefits seedlings, such as those of sugar maple, American beech, and eastern hemlock, that propagate well in poor light beneath the forest

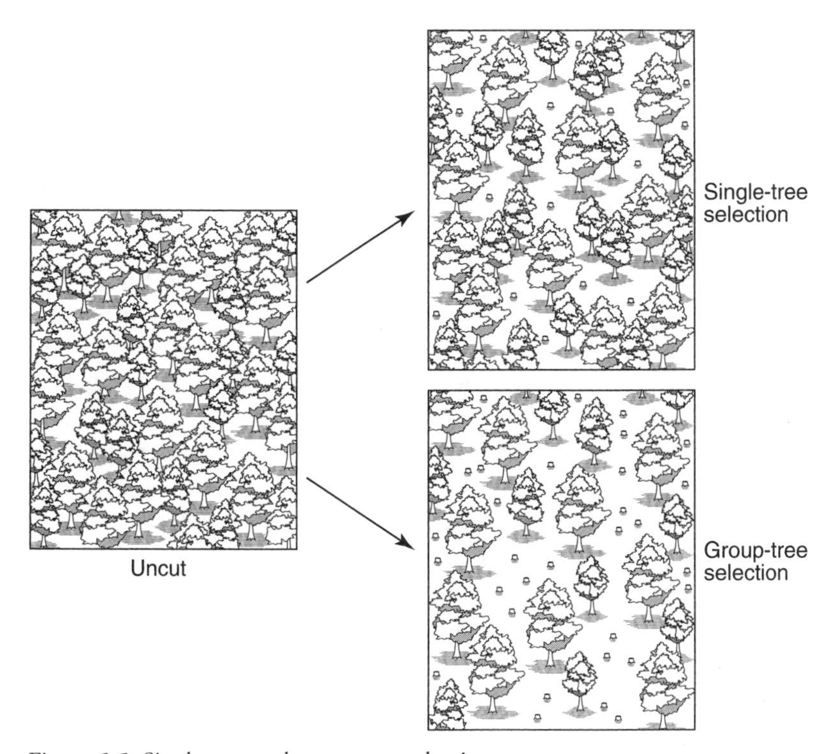


Figure 5.5. Single-tree and group-tree selection.

canopy. Group-tree selection creates larger openings in the canopy than single-tree selection and, thus, is better suited for seedlings that prefer partial shade, such as black cherry, tulip poplar, and white ash. Selective cutting is not recommended for timber management of southern bottomland forests or for oaks and hickories, which are shade-intolerant species.

Selective cutting is generally an acceptable practice for private landowners who wish to manage woodlots or other small forest stands to obtain income from timber without dramatically changing the structure and composition of flora and fauna in the stand. Both types of selective cutting allow a forest stand to be used for multiple purposes, such as timber production and aesthetics, while simultaneously helping to preserve the biological diversity associated with mature forest stands (Miller 1990; Thompson 1993).

From a wildlife perspective, removal of large oaks and hickories by selective cutting may be detrimental to species of birds and mammals that rely on mast as a source of food. However, in Illinois, selective cutting had no apparent effect on densities or survival of adult gray and fox squirrels (Nixon, Havera, and Hansen 1980). The effects of uneven-aged management on bird communities have not been given much attention. Both single-tree and group-tree selection are more beneficial to some bird species, such as hooded warblers, parula warblers, acadian flycatchers, and red-eyed vireos, than even-aged management, such as clear-cutting and shelterwood methods (Annand and Thompson 1997). In the Ozark–St. Francis National Forest in Arkansas, abundances of indigo buntings, eastern wood-pewees, and white-breasted nuthatches were most common in forested stands treated by both group-tree selection and understory removal than in untreated stands (Rodewald and Smith 1998). Single-tree selection may create favorable habitat for forest songbirds, like ovenbirds, wood thrushes, and American redstarts (Crawford, Hooper, and Titterington 1981). But bird species typical of mature forest, such as wood thrushes, ovenbirds, and worm-eating warblers, generally are more characteristic of unmanaged forests (Annand and Thompson 1997; Rodewald and Smith 1998).

### *Clear-Cutting Method*

Clear-cutting is the removal of all trees from a given area to produce a new stand of trees of similar size and age (Society of American Foresters 1981) (Figure 5.6). It is the most common method of even-aged management,

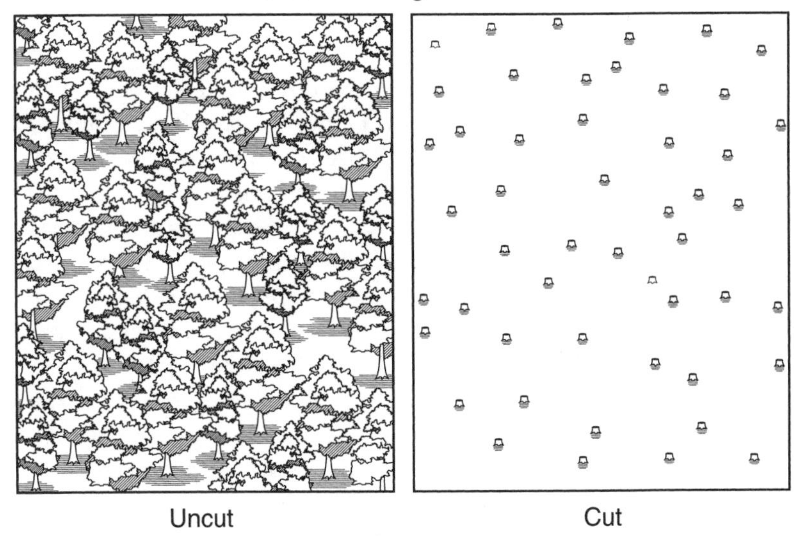


Figure 5.6. The clear-cutting method.

accounting for about two-thirds of the timber harvested annually in the United States and one-third of the harvest on national forests (Miller 1992). Because all sizes and species of trees are harvested, this method is cost and labor efficient. In general, all mature trees within an area of at least 1 hectare must be removed in order for the area to be termed a clear-cut stand (Hunter 1990). Like fire, clear-cutting creates a disturbance in the forest; but unlike fire, intensive or persistent clear-cutting can deplete important nutrients and minerals that have accumulated in the biomass of the forest (e.g., DellaSalla et al. 1995). These nutrients and minerals are recycled in a forest stand via ash produced by a fire, whereas they are virtually lost to the stand when timber is removed by clear-cutting.

Wildlife and forest managers probably know more about the effects of clear-cutting on plants and animals than they do about most other methods of forest management combined. Many tree species in the eastern deciduous forest, such as birches, red maple, black cherry, white ash, white pine, tulip poplar, hickory, swamp tupelo, and southern pines, regenerate favorably under this management practice (Society of American Foresters 1981; Lorimer 1990a). Oak and aspen also regenerate well after clear-cutting by sprouting from stumps and roots, respectively; this type of regeneration, termed coppicing, is used for pulpwood and fuelwood production in the eastern deciduous forest (Lorimer 1990a).

Clear-cutting can also be used to remove trees in a forest stand severely affected by disease or insect infestations.

The types of trees that regenerate after clear-cutting of a given forest stand are dependent largely on whether seedlings have germinated and dormant seeds are present in the soil before the cut. If seedlings and seeds are present before clear-cutting, species composition is less likely to change appreciably between pre- and postclear-cutting (Hunter 1990; Lorimer 1990a). On the other hand, if seedlings and dormant seeds are not available before tree removal, the tree species that regenerate will typically be those whose seeds are dispersed into the stand by wind or animals. Shade-intolerant tree species readily become established in recently clear-cut stands because these species, for example, maples, generally produce small seeds that are easily transported by winds. Unfortunately, our understanding of the long-term effects of clear-cutting on herbaceous plants, such as violets that provide beautiful wildflower displays in spring before canopy trees leaf out, is relatively limited and somewhat controversial. Some evidence suggests that clear-cutting of forty- to ninety-year-old and old-growth forest stands will result in irreparable damage to the diversity of understory herbaceous plants (Duffy and Meier 1992). On the other hand, very little difference was found in species composition or percent cover of herbaceous plants in older clear-cut stands (twenty years since clear-cutting) compared to mature stands (> 70 years since selective cutting) in central Appalachian stands (Gilliam, Turrill, and Adams 1995). If the recovery of herbaceous plants in managed forests is desired, the size of logged areas should be reduced to better mimic the size of gaps created naturally in the forest (Meier, Bratton, and Duffy 1995). To achieve this goal, selective cutting may be more preferable than clear-cutting, and log removal methods should be those that reduce damage to forest-floor vegetation and minimize soil erosion and compaction.

Size and shape of clear-cut stands are important considerations from an aesthetic perspective. Small, irregularly shaped clear-cut stands blend better into the landscape than large square-shaped or round clear-cut stands (Figure 5.7). Furthermore, size and age of clear-cut stands in the eastern deciduous forest are two major factors that affect the distribution and the abundance of wildlife (Conner, Via, and Pather 1979; Crawford, Hooper, and Titterington 1981; Yahner 1986a, 1989).

Large clear-cut stands may be especially suitable to some species that favor open habitats, like field sparrows or yellow-breasted chats (Crawford, Hooper, and Titterington 1981). In one recent study, the diversity

Figure 5.7. Aerial photograph of small, irregularly shaped clear-cut stands in the Allegheny National Forest, Pennsylvania. (Photograph by J. Timothy Kimmel, School of Forest Resources, Pennsylvania State University.)

of breeding bird species was higher in larger clear-cut stands compared to smaller ones (Rudnicky and Hunter 1993a). In contrast, large clear-cut stands can have a negative impact on forest wildlife that require large tracts of undisturbed habitat. Ovenbirds, red-cockaded woodpeckers, and red-eyed vireos may be representative forest species that are sensitive to disturbances created by large-scale clear-cutting in the eastern deciduous forest (Thompson et al. 1992; Irwin and Wigley 1993; Welsh and Healy 1993; Yahner 1993a, 1997a). Certain forest birds, such as ovenbirds, may not be affected negatively when territories are located in the vicinity (< 200 m) of small clear-cut stands (King, Griffin, and DeGraaf 1996). Furthermore, small clear-cut stands of less than 10 hectares provide valuable habitat for many songbirds (e.g., gray catbirds) and some mammals (e.g., snowshoe hares and white-tailed deer), which require dense, brushy vegetation for cover and food (Scott and Yahner 1989; Hughes and Fahey 1991; Yahner 1993a).

The effects of clear-cutting on forest wildlife are associated directly with the age of individual clear-cut stands. Older clear-cut stands typically have more breeding and wintering bird species as well as higher population abundances of these species compared to very young (e.g., < 5 years since cutting) clear-cut stands (Webb, Behrend, and Saisorn

1977; Yahner 1986a, 1993a). Young clear-cut stands are suitable to some bird species, such as ruffed grouse, eastern towhees, and gray catbirds, that require brushy habitat for foraging and breeding; these stands become less valuable to early-successional wildlife species as the stands begin to mature (e.g., twenty years since cutting) (McDonald, Palmer, and Storm 1994; Yahner 1997a; Lewis and Yahner 1999).

In part because of worldwide declines in amphibian populations (e.g., Wake 1991), several studies have been conducted addressing the effects of clear-cutting on terrestrial salamanders. The diversity and abundance of terrestrial salamanders are especially low in younger clear-cut stands (< 10 years old) than in older stands (> 50 years old) in southern Appalachian forests (Petranka, Eldridge, and Haley 1993). Forest management can also severely alter movements by salamanders, such as breeding migrations of the flatwoods salamander *(Ambystoma cingulatum)* when longleaf and slash pine savannahs were converted to plantations in north-central Florida (Means, Palis, and Baggett 1996). In central Pennsylvania, terrestrial salamander populations were at least 70 percent lower in clear-cut stands compared to older stands (> 70 years old) (R. H. Yahner, unpubl. data). In coastal British Columbia, clear-cutting reduced terrestrial salamander populations nearly 70 percent in old-growth forest stands (Dupuis, Smith, and Bunnell 1995). In the southern Blue Ridge Mountains of North Carolina, populations of terrestrial salamanders declined 30 to 50 percent within one year of clear-cutting, and virtually no salamanders were present two years after cutting (Ash 1997). Because of reduced leaf litter moisture in clear-cut stands, twenty to twenty-four years may need to elapse before salamander populations in these stands return to levels that occurred prior to cutting. Thus, in managed forest stands, every attempt should be made to preserve cool, moist microhabitats. This can be accomplished by retaining logs on forest floors, protecting spring seeps, preserving streamside buffers, and keeping some overstory as shade (Dupuis, Smith, and Bunnell 1995; Geer 1997). Clearly, additional research is warranted to fully understand the effects of forest management on long-term trends in terrestrial salamander populations (Ash and Bruce 1994).

The effects of clear-cutting on invertebrates (particularly arthropods), which are important food items for salamanders, birds, and mammals, also have been given attention in recent years (Greenberg and McGrane 1996). For example, ground-dwelling arthropods were more abundant in clear-cut stands than in mature forest stands, whereas the opposite occurred with shrub-dwelling arthropods (Van Horne and

Bader 1990). In contrast, another study found that invertebrate biomass was lower in clear-cut stands than in two-age or uncut forested stands (Wood, Duguay, and Nichols 1998).

A major reason why younger clear-cut stands are unsuitable for some wildlife is the absence of overstory trees, snags, and certain food resources. Clear-cut stands, for instance, are of limited value to wildlife, such as gray squirrels, that rely on mast, because mast-producing overstory trees are scarce or absent in clear-cut compared to uncut stands (Nixon, McClain, and Donohoe 1980). Without overstory trees or snags (Figure 5.8), clear-cut stands become less important as habitat to birds that require these features in which to nest, feed, or sing and to those species that require cavities for nest sites (Morrison et al. 1985; Dessecker and Yahner 1987; Yahner 1987, 1997a; Stribling, Smith, and Yahner 1990; Renken and Wiggers 1993; Machtans, Villard, and Hannon 1996). At least eighty-five species of birds in North America rely on tree cavities for nest sites (Scott et al. 1977). Many woodpeckers, like the pileated woodpecker, are dependent on snags as nest and foraging sites in mature stands throughout the eastern deciduous forest (Shackelford and Conner 1997). In relatively large clear-cut stands, red-tailed hawks often use snags as perches while hunting (Moorman and Chapman 1996). In fact, the retention of some overstory trees and snags can perhaps accelerate the regeneration of clear-cut stands, because birds that perch on these substrates may excrete seeds and thereby help disperse seeds within the stands (McClanahan and Wolfe 1993).

The retention of snags in managed forests is also valuable for the conservation of bats, which are important predators on insects (e.g., Kunz 1982). Forest bats, like the silver-haired bat, rely on large tree cavities (> 30 centimeters diameter breast height) for protection from inclement weather (Campbell, Hallett, and O'Connell 1996). Small mammals, such as white-footed mice, that rely on overstory trees and snags as foraging and nest sites often have lower populations in clear-cut stands than in mature stands (Yahner 1988b).

The pattern in which clear-cut stands are distributed in a forested landscape can influence their value to wildlife. Ruffed grouse, for example, require clear-cut stands of different ages in proximity to meet seasonal requirements for cover, food, and drumming sites (Gullion 1976). In northern Pennsylvania, snowshoe hares are more likely to be found in clear-cut stands that are within a reasonable distance (< 0.5 kilometers) of other clear-cut stands (Scott and Yahner 1989). Because Pennsylvania occurs within the southern fringe of the hare's geographic range, future

Figure 5.8. Snag with a cavity for wildlife. (Photograph by author.)

populations of this mammal may depend on the availability of suitable, nearby clear-cut stands.

Regardless of the size, shape, age, or distribution of clear-cut stands, the value of these stands as wildlife habitat may be diminished when the regeneration of tree species is poor. Stands may have inadequate regeneration because of low site quality, competition with ferns and grasses, or heavy deer browsing (Marquis 1981; Horsley and Marquis 1983). Poorly regenerated stands lack abundant vegetation near ground level, which provides cover, homesites, and food resources for songbirds and small mammals (Dessecker and Yahner 1987; Scott and Yahner 1989).

Clear-cutting affects not only terrestrial plants and animals but also aquatic life and forest soils. Like fire, clear-cutting of trees adjacent to streams can raise the water temperature and increase sediment (Brown and Krygier 1967; Society of American Foresters 1981). Hence, buffer strips of trees (30 to 75 meters wide) should be left standing along streams to reduce the effects of tree removal on aquatic habitat (Brazier and Brown 1973; Spurr and Barnes 1980). Furthermore, clear-cutting on steep slopes should be avoided because of potential problems with soil erosion, unless the area is adequately reseeded or replanted immediately after clear-cutting to minimize soil loss and ensure rapid regeneration of the stand (Lorimer 1990a; Miller 1992). Repeated clear-cutting on a site can cause considerable compacting of the forest soil by heavy logging equipment, thereby reducing the productivity of the site for years to come (Miller 1990; but see Donnelly, Shane, and Yawney 1991). On a localized scale, clear-cutting of a forest stand may superficially resemble a fire in terms of its effects on vegetation (DellaSalla et al. 1995). But unlike a fire, intensive or persistent logging can have a major impact by depleting essential nutrients, minerals, and elements that have been retained in the forest biomass through many decades or centuries.

That forest clear-cutting in the past has resulted in long-term effects on the distribution and the abundance of plant and animal life in the eastern deciduous forest is certainly plausible, but we cannot be certain about these effects at this time (Johnson, Ford, and Hale 1993). We would have difficulty, for instance, in separating the effects on the extinction of wildlife species of overexploitation by uncontrolled hunting versus habitat changes due to clear-cutting. We cannot help but stop and wonder, however, what impacts extensive deforestation via clear-cutting in the latter part of the nineteenth century and in the early decades of the twentieth century had on current patterns of plant and animal species in today's forest.

*Shelterwood Method*

In contrast to clear-cutting, the shelterwood method of timber harvesting involves a series of cuttings over a period of about ten years (Lorimer 1990a). In the first cutting, usually 30 to 50 percent of the mature trees are selectively harvested; diseased trees also may be removed (Society of American Foresters 1981; Patton 1992). Once seedlings become established, a second cutting may be applied to remove about 25 percent of the mature trees (Figure 5.9). These initial cuttings permit better light penetration, improve vigor and seed production of the remaining trees, and allow seedlings to get a good start beneath the shelter of mature trees. A final cutting removes the remaining trees, permitting the seedlings to grow to maturity, usually resulting in an even-aged stand (Society of American Foresters 1981; Miller 1990).

The shelterwood method of harvesting is used throughout the eastern deciduous forest. It is particularly recommended in small stands containing only a few trees species (Cutter, Renwick, and Renwick 1991) and is probably the best way to regenerate oaks (Society of American Foresters 1981). The shelterwood method, however, takes more expertise and is more costly than clear-cutting (Miller 1992). Like the effects of clear-cutting, the effects of the shelterwood method on wildlife can vary widely with species. In northern hardwoods, for in-

Figure 5.9. Shelterwood cut in central Pennsylvania. (Photograph by Colleen A. DeLong, School of Forest Resources, Pennsylvania State University)

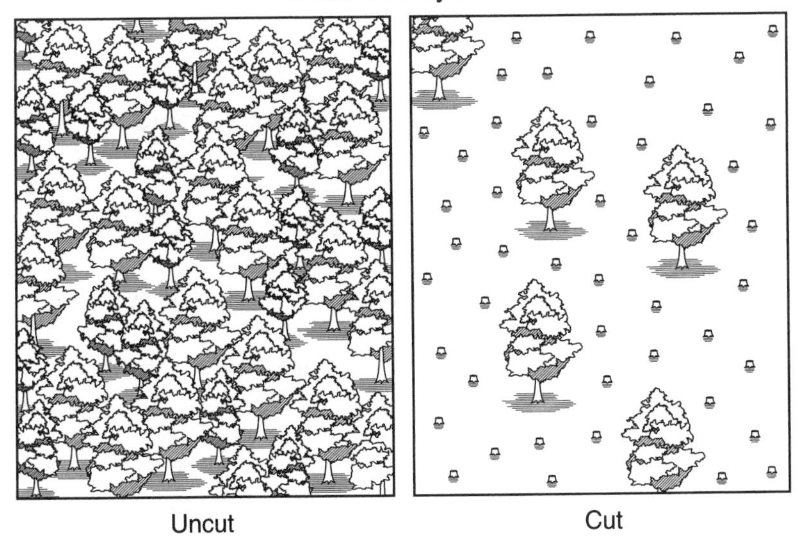


Figure 5.10. The seed-tree method.

stance, removal of 25 to 100 percent of the timber by shelterwood and other timber-harvesting methods had no effect on abundances of eleven species, a positive effect on eight species, and a negative effect on seven species of breeding birds (Webb, Behrend, and Saisorn 1977).

### *Seed-Tree Method*

The traditional clear-cutting method can be modified to better ensure regeneration, enhance aesthetics, and provide valuable habitat by retaining a certain number of trees in a stand after harvest. The seed-tree method of timber harvesting, for example, leaves a few mature, seed-producing trees in a given area after most trees in the stand are removed (Society of American Foresters 1981; Lorimer 1990a) (Figure 5.10). Usually at least five to ten trees per hectare are left singly or in a group to facilitate the regeneration of an even-aged stand (Miller 1990). Once the new trees become established in the stand, the seed trees may be harvested. The seed-tree method usually is not applied in northern hardwood, oak-hickory, or southern bottomland hardwood forest regions, but it has some value in regenerating southern pine species, such as loblolly pine (Society of American Foresters 1981). In central Appalachian hardwoods, trees retained in seed-tree stands provide cavities for nesting birds, such as eastern bluebirds (Crawford, Hooper, and Titterington 1981).

*Some Alternative Methods to Traditional Even-Aged Management*

Some forest-management agencies have adopted practices similar to the seed-tree method to replace traditional clear-cutting. In central Pennsylvania, the Pennsylvania Game Commission has retained about fifteen to twenty trees per hectare after even-aged management of oak stands for ruffed grouse habitat (Figure 5.11). These residual trees provide valuable perch sites for birds, like Baltimore orioles and cedar waxwings (Yahner 1993a; 1997a). In Minnesota, the retention of residual patches, consisting of fifteen or so closely spaced trees in aspen clear-cut stands, was valuable to forest birds, like veeries, ovenbirds, and black-throated green warblers (Merrill, Cuthbert, and Oehlert 1998).

A method used by the Pennsylvania Bureau of Forestry since 1992 to replace traditional clear-cutting on state forests is termed "even-aged reproduction stands with reservation guidelines," or simply the even-aged reproduction method (Pennsylvania Bureau of Forestry 1991; Boardman 1977) (Figure 5.12). About twelve trees and shrubs are retained per hectare in each even-aged reproduction stand to maintain vertical struc-

Figure 5.11. Residual trees left in an even-aged oak stand in central Pennsylvania. (Photograph by author.)

Figure 5.12. Residual trees left in an even-aged reproduction stand in central Pennsylvania. (Photograph by Katharine L. Derge and by author, School of Forest Resources, Pennsylvania State University.)

ture in the vegetation, increase aesthetics, and enhance floral diversity. Thus, these stands sometimes are referred to as "diversity cuts."

In central Pennsylvania, total abundance of all bird species combined in even-aged reproduction stands was similar to that in clear-cut stands with some residual snags (Boardman 1997; Boardman and Yahner 1999). However, abundance of black-capped chickadees was higher in the even-aged reproduction stands than in the clear-cut stands with snags, but abundances of both blue jays and great crested flycatchers were lower in the even-aged reproduction stands compared to the clear-cut stands. Residual trees in even-aged reproduction stands are important perching and singing sites for many bird species, such as chipping sparrows and scarlet tanagers; these trees are also used as foraging and nesting sites by white-footed mice. Unharvested mast-producing trees, like oaks in even-aged reproduction stands, provide food resources for many wildlife species (see chapter 3).

Another alternative to traditional methods of even-aged management is the two-age cutting method, which is sometimes referred to as deferment cutting (Smith, Lamson, and Miller 1989; Miller, Johnson, and Baumgras 1997) (Figure 5.13). The two-age cutting method has been used in recent years on public and private forestland in several eastern states. This cutting method is valuable in regenerating high-quality

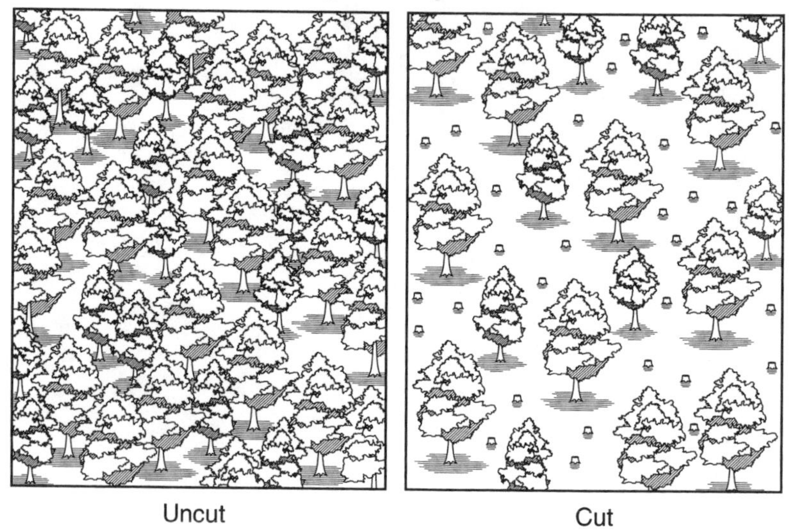


Figure 5.13. Two-age cutting method.

hardwoods, particularly shade-tolerant species, and in creating a stand with two ages of woody vegetation for both wildlife habitat and non-timber benefits. With a two-age cut, thirty to forty-nine residual trees per hectare are retained, which is about three to ten times the number of residual trees left with some other methods of forest management, e.g., seed-tree or even-aged reproduction methods. Once the woody vegetation has regenerated sufficiently after about forty to eighty years, the residual trees are then harvested. As with seed-tree and even-aged reproduction stands, mast trees can be left unharvested to provide food for wildlife.

A study of two-age cuts in West Virginia has shown that most (89 percent) residual trees had survived ten years from the time the stands were harvested; furthermore, regeneration of the new age class was very successful (Miller, Wood, and Nichols 1995). When compared to clear-cut stands of similar age since harvest, total abundance of all breeding bird species combined was considerably higher in the two-age cuts. However, bird species composition in both clear-cut stands and two-age cuts was relatively similar. In another study of two-age cuts and clear-cut stands that had been harvested about fifteen years earlier, total abundance of all bird species combined was similar in the two stand types (Duguay 1997). Abundance of the most common breeding species, such as red-eyed vireos, ovenbirds, and brown-headed cowbirds, did not differ between the two types of stands. In contrast, bird species that nested in shrubs or foraged at ground level were more abundant in two-age cuts than in clear-cut stands.

### *Thinning*

Select trees in a forest stand may be harvested well in advance of a scheduled timber harvest by a process termed thinning (Patton 1992). The amount of timber removed from a given site may vary from 10 to 50 percent (Crawford and Titterington 1979). If the thinned trees are usable as timber, the practice is called commercial thinning. This thinning may help increase the growth of the remaining trees, especially in a dense stand of unevenly spaced trees. If thinned trees are unusable because of damage from disease, fire, or some other factor, the thinning is called a salvage cut. Thinning overmature trees benefits some bird species of the eastern deciduous forest, such as hooded warblers and northern cardinals (Crawford, Hooper, and Titterington 1981). Retaining dead trees, which are used for foraging and as homesites for birds and other wildlife, can have considerable long-term benefits for the eastern deciduous forest (Yahner 1987; Stribling, Smith, and Yahner 1990).

### *Some Forest-Management Policies and Controversies*

During the past century, forests in North America generally have been managed either as plantations for timber production or as preserves for wildlife and recreational benefits (Spies et al. 1991). In the 1950s and 1960s, however, the management of forests for multiple values, termed multiple-use management, was initiated on national forests by the U.S. Forest Service, but natural resource managers in those decades unfortunately had limited ecological understanding of the forest ecosystem to implement this approach (Franklin 1997). In the 1980s, forest management entered into an era of "new forestry," which was intended to expand an ecological understanding of forests and also to ensure that management practices were both ecologically and socially conscious (Gillis 1990). New forestry described philosophies of managing forests that gave attention to other values of the forest rather than focusing solely or principally on timber production (Spies et al. 1991; DeBell and Curtis 1993). Moreover, new forestry had a major objective of maintaining biological diversity (see chapter 8) in managed forest stands and assessing management activities at the landscape level and over longer time periods than previously considered by earlier forest-management practices (Franklin 1990).

In the late 1980s through the early 1990s, the Forest Service implemented its "new perspectives," which challenged traditional forest-management practices while embracing sound ecological concepts and social considerations into its policies. Today, most forests are viewed more holistically—not merely as a source of timber but also as an opportunity for wildlife and aesthetic values. This broader approach to forest management is relatively recent and is called ecosystem management (Grumbine 1994; Franklin 1997).

Some natural resource managers may contend that new forestry and ecosystem management have been done successfully for years under multiple-use forest management (O'Keefe 1990). In fact, we can go back to early in the nineteenth century and find the term new forestry being used in another context to describe management practices for tree plantations in Germany, England, and North America (Spies et al. 1991). Furthermore, we can look back a few decades to the late 1940s with the publication of "A Sand County Almanac" by Aldo Leopold, in which he formulated the idea of a "land ethic" (Leopold 1949). This land ethic called for the stewardship of lands to conserve biodiversity, commodities, and amenities, which is the basis for ecosystem management today (Knight 1996). As we will see in chapter 8, ecosystem management has a rightful place in natural resource conservation, particularly as it relates to the conservation of forest biodiversity.

Natural resource managers must work cooperatively to find ways of managing forests that are compatible with multiple goals (Eastman et al. 1991; Ticknor 1993). An excellent example of how multiple goals can be achieved cooperatively is the integration of conservation efforts for Neotropical migratory songbirds and ecosystem management projects on lands under the jurisdiction of the National Wildlife Refuge System and the U.S. Forest Service (Clark 1993; Finch et al. 1993). Unfortunately, cooperative efforts and the use of expertise to manage forests for multiple purposes will continue to be constrained by limited budgets, time, and management options.

Of the various methods of timber harvesting, the clear-cutting method has been the most controversial. Clear-cutting was a common silvicultural method in national forests from the 1950s to the early 1970s, but the National Forest Management Act of 1976 helped set limits on timber harvest in these forests (Miller 1990). Some opposition to clear-cutting rose, in part, because of a greater awareness of the need to manage the forest from an ecosystem perspective. For instance, clear-cutting of the last remnants of old-growth forest in the Pacific Northwest with little regard to the overall impact of the practice on this valuable ecosystem became a major conservation issue over the past several decades (Noss 1991). As a result, in 1992, the U.S. Forest Service declared that clear-cutting would no longer be the standard way of harvesting timber on national forests, possibly reducing its use by as much as 70 percent (Hill 1992). Some environmentalists feel that clear-cutting in national forests should be banned altogether, whereas others feel that clear-cutting can be used in forest management under appropriate ecological, economic, and social conditions. This forest-management practice may have a rightful place in forest management and be used under certain circumstances, such as to manage habitat in national forests for threatened, endangered, or sensitive species of wildlife that require early-successional habitats.

As discussed earlier in this chapter, many regard clear-cutting as a satisfactory management practice for maintaining a diversity of wildlife in the eastern deciduous forest landscape, particularly when cuts are limited in size (Gullion 1990; Welsh and Healy 1993; Yahner 1993a, 1997a) and when seed or residual trees are retained (Yahner 1993a, 1997a; Boardman 1997; Boardman and Yahner 1999). Today, forest management clearly is headed in the direction of practices that make a concerted effort to retain individual trees, snags, logs, and small patches of trees within the managed stand and landscape in an effort to meet ecological objectives, for example, to maintain biological integrity. These practices, broadly referred to as variable retention harvest systems (Franklin et al. 1997), hold considerable promise for the sound management of the eastern deciduous forest from an ecosystem perspective. Moreover, larger managed stands, such as those created by traditional clear-cutting devoid of residual trees, will and, in my opinion, should be less common in the future because of unfavorable aesthetics, limited value to many wildlife species, and possible impacts on soil erosion (Lorimer 1990a). In addition, the impact of forest-management practices on the ecosystem can be reduced if time between harvests (e.g., rotation) can be extended whenever possible beyond that currently in use (Franklin et al. 1997). Conversion of large, relatively mature forested stands into smaller tracts via clear-cutting, agriculture, or other land uses and its consequences for the distribution and the abundance of forest plants and animals are discussed further in the next three chapters.

# 6. FOREST FRAGMENTATION

Forest fragmentation results when a large, relatively mature forested stand is converted into one or more smaller forested tracts by human land uses, such as agriculture, urbanization, or timber harvesting. Fragmentation not only reduces the amount of forest but also isolates the remaining forested tracts from one another because of the intervening land use (Harris and Silva-Lopez 1992). Thus, we could visualize forest fragmentation as a process that leads to islands of forested tracts within a sea or landscape of agriculture, urbanization, or younger stands of managed forests (Figure 6.1).

Forest fragmentation and its consequences for forest biota have become major concerns within the wildlife and forestry professions, but only since the late 1970s (Yahner 1995a). That they were not concerns earlier is perhaps because the biological principles from which patterns of the distribution and the abundance of wildlife can be explained were only developed in the late 1960s. One important principle that emerged for vertebrates in particular is the species-area curve (Figure 6.2) (MacArthur and Wilson 1967).

The species-area curve is based on the island-biogeography theory, which was used initially to explain distributional patterns of species on oceanic islands of different sizes. This curve predicts that a greater number of species (species richness) will be found on a larger island (habitat) than on a smaller island (habitat). For example, a large, isolated oceanic island typically contains more species of organisms than a small, isolated island. The relationship between island size and number of species is also influenced by the degree of isolation, which affects organisms' rates

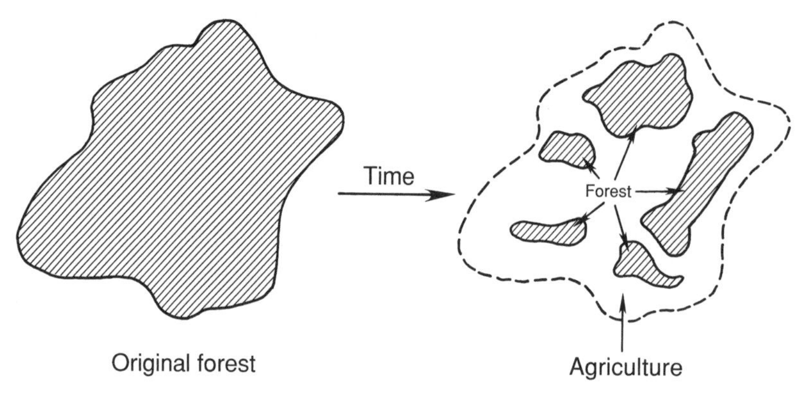


Figure 6.1. An original forest stand fragmented over time into smaller forested tracts by an intervening land use, in this case agriculture.

of colonization of an island (Figure 6.3). Furthermore, greater rates of extinction generally are expected on a smaller island because of smaller population sizes; by chance alone, a smaller population has a greater likelihood of going extinct than a larger population. Hence, a small, isolated island typically contains fewer species than a larger, less isolated island. The effect of size and isolation is greater on less mobile species, such as small mammals and wind-dispersed plants, than on more mobile species, such as birds and large mammals (Harris 1984).

Although the size of islands per se is a major factor determining the number of species occurring on a given island, other related factors can be important. For example, islands usually are not homogeneous but rather contain a variety of habitat conditions or environments. Larger islands, in particular, would be expected to be more heterogeneous than smaller islands, thereby enabling a higher number of wildlife species to find suitable food, cover, and other requisites for survival.

Before the theory of island biogeography can be used to assess the effect of forest fragmentation on the distribution and abundance of forest biota, at least two major differences between an oceanic and a terrestrial island must be appreciated. First, an oceanic island often is older geologically, having been formed thousands of years ago by volcanic activity or corals (Harris 1984; Hunter 1990). At the other extreme, a large forested tract can be converted to a series of terrestrial islands within a matter of only a few months or years by extensive clear-cutting or agriculture. Second, an oceanic island is surrounded by water, which may serve as an impassable, permanent barrier to movements of many terrestrial organisms, except those that are aerial or wind-borne. In contrast, a clear-cut stand or an agricultural field adjacent to a terrestrial forested island is less insurmountable than water to movements by most terrestrial organisms. Although public lands, such as national parks, are not true isolates, detailed examinations of long-term trends in wildlife have documented rates of extinction predicted by island-biogeography theory (Newark 1995). For instance, in fourteen western national parks, extinction rates increased as the size of parks decreased.

## History of Forest Fragmentation

Forest or habitat fragmentation has been described as the most serious threat to biological diversity, or biodiversity, and is a primary cause of accelerated rates of extinctions of plants and animals worldwide (Wilcox and Murphy 1985). Hence, a major challenge facing society and natural resource managers alike is to maintain the world's biodiversity despite a continual loss of available forested habitat. Concerns for loss of forested

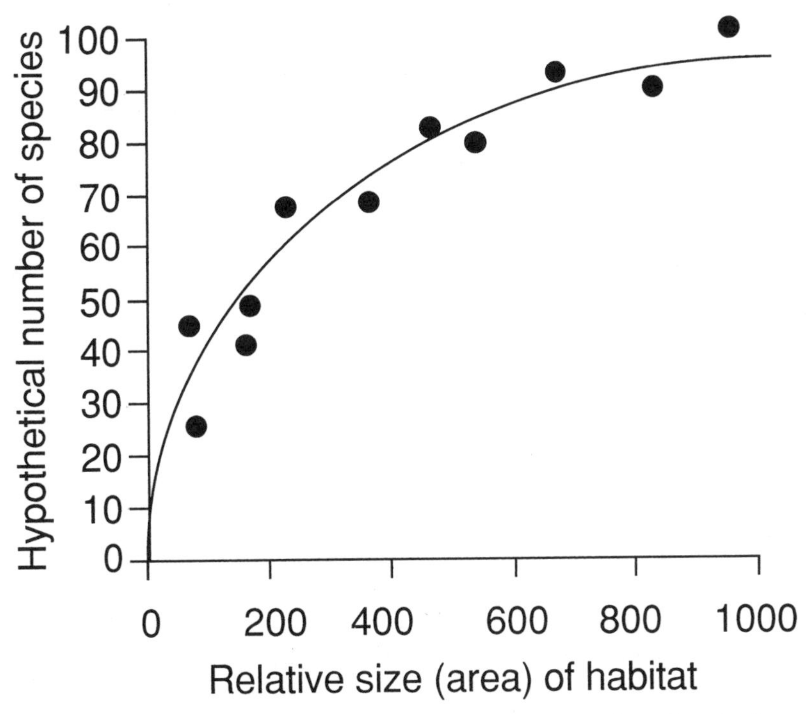


Figure 6.2. A hypothetical species-area curve giving the number of species in relation to size (area) of habitat. Each dot represents a habitat of a given size, containing a certain number of species.

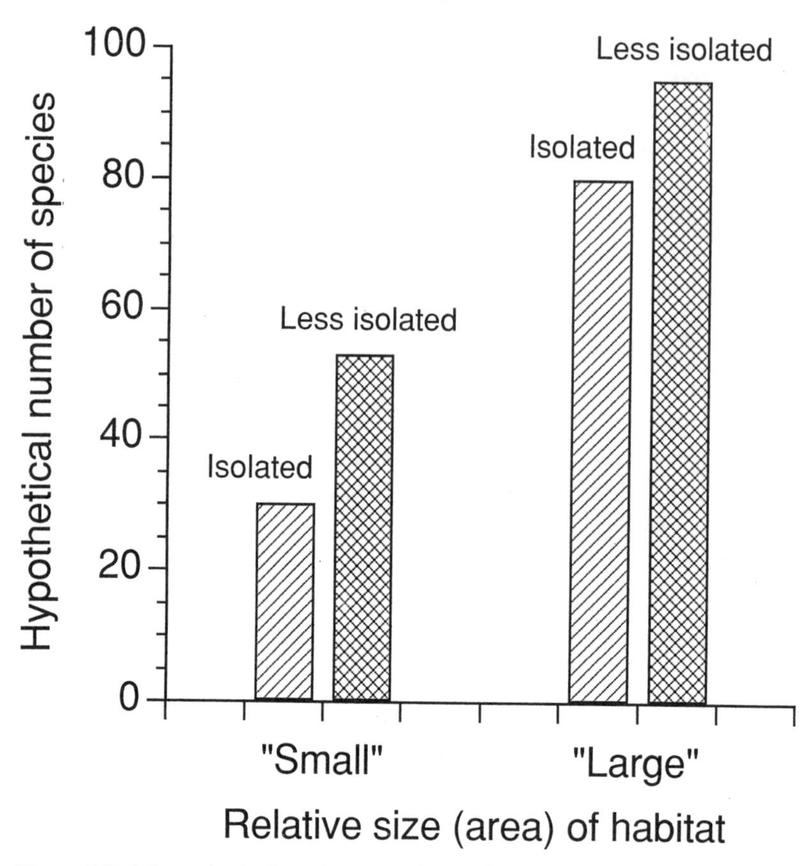


Figure 6.3. A hypothetical species-area relationship in habitats that are isolated versus those that are less isolated from similar habitats.

habitat and biodiversity worldwide, along with the issue of global climatic change, were main agenda items at the historic Earth Summit in Rio de Janeiro in 1992, which was attended by 178 nations (Nicholls 1992; Shabecoff 1992).

Forest fragmentation in the tropics is a major conservation issue that has remained in the forefront in contemporary media and scientific publications, and all sectors of society are beginning to understand the ramifications of this issue for forest biodiversity. Reasons for this concern are clear. About 7 percent of the earth's land surface is tropical forest, yet nearly 50 percent of the original forest has been lost to timber harvesting, cattle grazing, fuelwood, and farming (Miller 1990). Each year, about 1 percent of this forest is deforested and another 1 percent degraded, which represents more than 150,000 square kilometers of forest per year (Myers 1988; Miller 1990). An alarming fact is that at least 50 to 80 percent of the world's species of plants and animals occur in these tropical forests; thus, many biologists predict mass extinctions of tropical plants and animals during the next couple of decades.

Forest fragmentation in temperate latitudes, such as in the eastern deciduous forest, has not received the same level of attention given to tropical deforestation (Yahner 1995a). Until recently, deforestation in the eastern deciduous forest had been viewed with less alarm by scientists and society perhaps for two reasons. First, compared to tropical species, temperate species generally have greater population densities and broader geographic ranges (Wilcove, McLellan, and Dobson 1986). Therefore, if two large expanses of forest of comparable size were deforested, one in the tropics and the other in the eastern deciduous forest, we would expect that more species would be lost in the tropical forest. Second, forest fragmentation in temperate latitudes may have been largely hidden from view because much of it occurred many decades ago, well before it became a conservation issue to most people. As we saw in chapter 2, fragmentation of the eastern deciduous forest began in the seventeenth century with the arrival of Europeans and reached a peak in the late nineteenth and early twentieth centuries (Wilcove, McLellan, and Dobson 1986). In fact, the amount of forested land, which in the 1500s spanned about 400 million hectares in the forty-eight contiguous states (Harrington 1991), was reduced to about 188 million hectares (a decline of 53 percent) by the 1920s. Today, about 95 to 97 percent of the remaining old-growth forest, consisting of virgin stands, is gone from those states. Virtually no old-growth forest remains in the eastern United States; old-growth forest in the lower forty-eight states is principally confined to the Pacific Northwest (Figure 6.4) (Miller 1992).

Much of the current forest in parts of the eastern United States is relatively immature. For example, only 22 percent of central Maryland was forested in the 1980s, and most of the forest is in early- or mid-successional stages of growth (Whitcomb et al. 1981). About 59 percent of Pennsylvania is forested today, consisting principally of stands that have regenerated since the 1920s and 1930s. In addition, about 40 percent of the deciduous forest in the East consists of small, isolated woodlots in suburbia and farmlands (Terbough 1989). Today's younger and highly fragmented forest cannot be expected to support the same variety of plant and animal species as the pristine, old-growth forest that once covered the East.

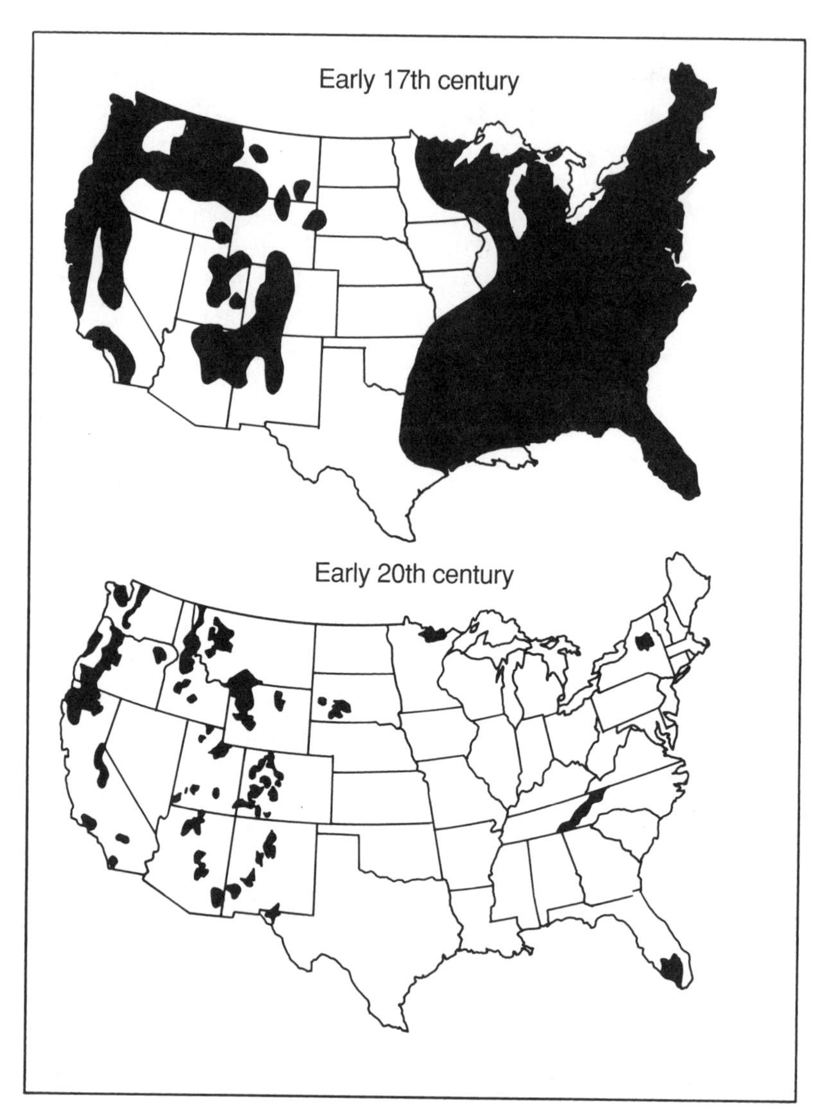


Figure 6.4. Comparison of the distribution of old-growth forest in the lower forty-eight states in the early seventeenth century and the early twentieth century. (Modified from Miller 1992.)

In many areas of the eastern United States, relatively extensive tracts of forest in parks and nature reserves have become "green islands" surrounded by dissimilar habitats or land uses, making them less suitable for wildlife (Harris 1984). For instance, in the Great Smoky Mountains National Park, recent land-use changes and road development along park boundaries are predicted to have a major impact on wildlife. Extensive roadways not only alter the suitability of wildlife habitat but can affect dispersal and mortality of wildlife species that occupy forested habitats within and adjacent to the park (Ambrose and Bratton 1990). Roads are more likely to have a negative influence on wildlife species that use forest-interior habitat, are sensitive to human presence, occur at low densities, avoid crossing roads, and are wide ranging (Harris and Gallagher 1989; Schonewald-Cox and Buechner 1992). Roads along forest edges, for example, are major barriers to movements by amphibians (Gibbs 1998). Thus, roadless areas should be preserved for the benefit of wildlife species requiring undisturbed habitat for survival, particularly in large tracts of wilderness (The Wildlife Society 1998); furthermore, efforts should be made to decommission old roads, upgrade selected roads to minimize ecological impacts to wildlife, and restore forests damaged by the existing network of roads.

## Forest Fragmentation and Wildlife Extinctions

Extinction and endangerment of several wildlife species in the eastern deciduous forest have been attributed partially to forest fragmentation that continued into the twentieth century. For instance, both fragmentation and uncontrolled hunting pressure led to the extinction of the passenger pigeon in 1914, which was a species that perhaps once numbered two billion in the eastern deciduous forest (Bolen and Robinson 1995). The passenger pigeon relied on oak and beech forests for roosting, nesting, and food. Another bird, the ivory-billed woodpecker, which was the largest woodpecker in the United States, was extirpated during this century concurrent with the destruction of swamp forests throughout the Southeast; a few still may exist in Cuba (Gill 1990).

The red wolf once roamed much of the southeastern forest from Florida to southern Pennsylvania and west to central Texas (Paradiso and Nowak 1982). Conversion of the deciduous forest to farmland in the 1950s and hybridization between red wolves and coyotes were two factors responsible for the near demise of red wolves in the wild. Captive breeding programs for this wolf that began in 1973 and attempts to reestablish it in North Carolina starting in the late 1980s have progressed well (Phillips and Parker 1988; Gilbreath 1998). As a result of these conservation efforts, there is hope that the red wolf once again may return successfully to some forested fragments in the eastern United States.

Timber removal in river bottoms and lowlands in the eastern United States coupled with unregulated hunting nearly extirpated our most beautiful species of waterfowl, the wood duck, in the early twentieth century (Bellrose 1976). However, with the passage of the Migratory Bird Treaty Act of 1918, the wood duck gained protection through controlled harvest regulations. Numbers of wood ducks subsequently increased to the point that it is now one of the most common waterfowl species in the eastern deciduous forest.

Even today, large numbers of invertebrate species (particularly insects) are or potentially will be lost, as ancient (old-growth) forests in the tropics and our Pacific Northwest are removed forever. Forest fragmentation probably has caused the direct elimination of plant populations worldwide for many centuries (Ledig 1992). Unfortunately, in the eastern deciduous forest, we have little knowledge of how many species of lesser known vertebrates, invertebrates, and plants have become extirpated because of forest fragmentation that has been ongoing since the seventeenth century (Hafernik 1992).

## Fragmentation and Forest Wildlife

The effects of forest fragmentation on wildlife in the eastern deciduous forest have been best studied in birds. Since the 1970s, several field studies have indicated that the size of forested tracts combined with the extent to which these tracts are isolated can negatively affect the diversity and the distribution of breeding bird species (Robbins, Dawson, and Dowell 1989; Wiedenfeld, Messick, and James 1992). In New Jersey, for instance, the number of breeding bird species decreased as the size of forested tracts within agricultural and residential landscapes became smaller; these tracts ranged from 0.01 to 24 hectares (Galli, Leck, and Forman 1976) (see also Figure 6.2). In South Carolina, species richness of Neotropical migrants expanded as the size of hardwood stands surrounded by either agricultural or pine habitat expanded (Kilgo et al. 1997). Moreover, the relationship between size of uncut forested tracts and number of bird species can vary seasonally. In south central Pennsylvania, for example, the number of bird species during spring migration and the breeding season increased as tract size increased, but the correlation between number of species and tract size did not occur during fall migration or winter (G. S. Keller and R. H. Yahner, Pennsylvania State University, unpubl. data).

In a forest-dominated landscape of central Pennsylvania, fragmentation of the forest into small stands of different ages as a result of even-aged management provided suitable habitat for a variety of wildlife species. Here, a mosaic of small (1 hectare) even-aged plots were created by three cutting cycles (1976–77, 1980–81, and 1985–87) in a localized area (Figure 6.5). These plots not only created numerous "activity centers" (Gullion 1976) for ruffed grouse but also supported a high diversity of songbird species. Golden-winged warblers, chestnut-sided warblers, and other songbirds would not occur in this area were it not for the presence of early-successional forested plots managed for grouse habitat (Yahner 1986a, 1993a, 1997a). Yet despite the fact that fragmentation of a forest into small even-aged plots may increase wildlife diversity on a local basis, it may not necessarily result in an overall net increase in the regional diversity of wildlife (Murphy 1989).

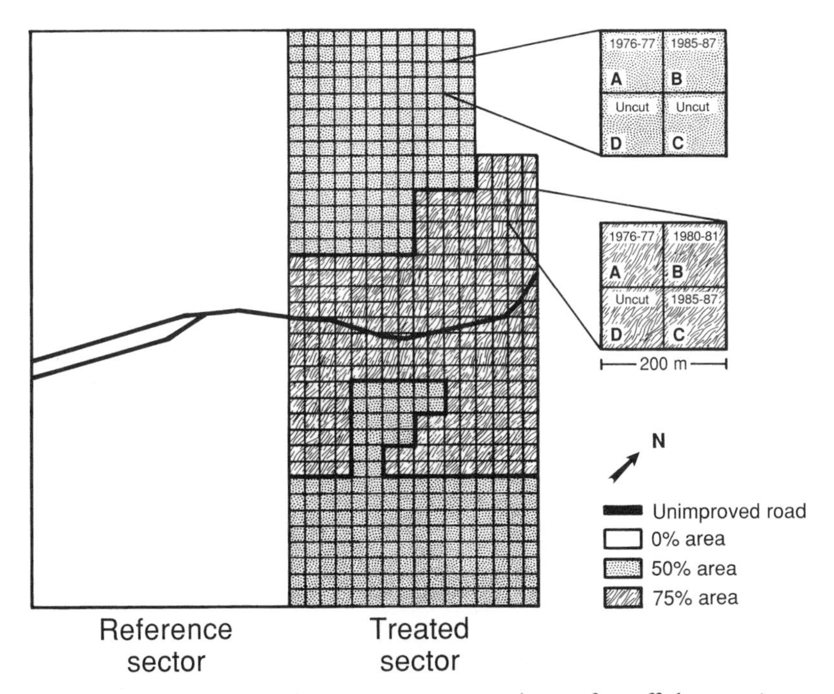


Figure 6.5. An intensive habitat management study area for ruffed grouse in Pennsylvania, known as the "Barrens." (Modified from Yahner 1992.)

Natural resource managers obviously should be concerned about forest fragmentation's long-term impacts on wildlife. In landscapes altered by land uses, such as the residential development in the New Jersey study cited earlier, the distribution and the abundance of forest wildlife is likely to be affected permanently. However, the impact of forest fragmentation on wildlife may not necessarily be long term. Consider the central Pennsylvania example given earlier, which dealt with a forested landscape fragmented by even-aged management. Here, the distribution of mature forest birds, such as red-eyed vireos, was reduced in the short term by the creation of recent even-aged plots. But after ten to twenty years of plant succession, these plots gradually became more suitable habitat for red-eyed vireos, ovenbirds, and other forest bird species, whereas some early successional bird species, e.g., common yellowthroats and field sparrows, exhibited marked declines (Yahner 1993a, 1997a). Thus, because silvicultural practices such as clear-cutting may have a relatively short-term and less destructive effect on the forested landscape and its biota compared to urbanization or other landscape changes (Yahner 1997b), some have suggested that these landscapes be referred to as variegated rather than fragmented (McIntyre and Barrett 1992). Most scientists would agree that any fragmented (or variegated) forested landscape, even if given enough time, will never revert back to the old-growth stands that existed in eastern North America from the sixteenth through the eighteenth centuries (Lorimer and Frelich 1994).

Today, there is a growing concern about what I label "reverse forest fragmentation" (Yahner 1995a). In recent decades in some areas of the eastern deciduous forest, particularly in the Northeast, early-successional habitats are reverting back to secondary or mature forest at an unprecedented rate. This forest maturation will have some important implications for long-term population trends in certain wildlife species. For instance, with forest maturation in New England during the past three decades, species dwelling in semi-open or edge habitats, such as eastern towhees, have decreased rapidly (DeGraaf, Askins, and Healy 1993; Hagan 1993). Maturation of forests in the Northeast is probably contributing also to declines in populations of American woodcock, New England cottontails, and several other species of songbirds (Coulter and Baird 1982; Thompson et al. 1992; Wiedenfeld, Messick, and James 1992; Litvaitis 1993). In another case, forest maturation via the natural conversion of deciduous stands to eastern hemlock stands has adversely affected the distribution of the threatened Appalachian woodrat in Pennsylvania (Balcom and Yahner 1996). In a reverse situation, the creation of early-successional habitat associated with wetlands, via cutting or burning of woody vegetation, has been detrimental to the conservation of populations of the eastern massasauga rattlesnake, which is a species of concern in several areas of the eastern deciduous forest, such as Ontario and New York (Johnson and Leopold 1998). Therefore, both forest fragmentation and maturation are landscape processes that natural resource managers must consider to better understand long-term trends in wildlife populations in the eastern deciduous forest (Yahner 1995a).

### *Fragmentation and Long-Term Trends in Bird Populations*

In the eastern deciduous forest, about 60 to 89 percent of the breeding bird species, including many species of warblers, vireos, and tanagers, are Neotropical migrants that overwinter in the forests of Central and South America. Many of these songbirds play key ecological roles in the forest as predators of insect pests and are important from an aesthetic perspective to many people who participate in outdoor activities, such as birding. A major study in the late 1980s reported that populations of approximately 70 percent of the migratory species in the eastern deciduous forest have experienced declines (Robbins, Dawson, and Dowell 1989). A second study found that populations of twenty-five (81 percent) of thirty-one migratory species in eastern North America may have been reduced from 1978 to 1987 compared to the period from 1966 to 1978 (Askins, Lynch, and Greenberg 1990). A third study documented that sixteen species of forest-dwelling Neotropical migrants, such as eastern wood-pewees and Kentucky warblers, declined over the period from 1966 to 1991 (Peterjohn and Sauer 1994).

Some other analyses of bird populations, however, have presented a different picture of long-term trends in bird populations, suggesting that declining populations trends described in the studies just mentioned were oversimplified (Yahner 1995a). For instance, based on information obtained by breeding bird surveys conducted by the U.S. Fish and Wildlife Service during 1966 to 1990, only nine (14 percent) of sixty-five species showed significant population declines, whereas thirty-two (49 percent) species actually increased in abundance (Wiedenfeld, Messick, and James 1992). Wood thrush and gray catbird populations were among those that declined, and ovenbird and black-capped chickadee populations were representative of those that increased. Interestingly, an analysis of these surveys found that many species were prone to exhibit declines or increases at certain "hot spots" in the eastern deciduous forest. As an example, populations of more than two-thirds of the sixty-five species decreased in four major hot spots: Cumberland Plateau, Blue Ridge Mountains, Mississippi Alluvial Plain, and Adirondack Mountains.

In another recent study, populations of twenty-six species of woodland warblers in eastern and central North America were analyzed from breeding bird surveys conducted from 1966 to 1992 (James, McCullough, and Wiedenfeld 1996). This study confirmed that changes in population numbers tended to occur mainly in the early 1970s rather than later, with four species showing definite population declines and nine exhibiting major increases in numbers. Examples of declining species were prairie and golden-winged warblers, yellow-breasted chats, and American redstarts; some increasing species were ovenbirds and mourning, blue-winged, and magnolia warblers. A bottom line to this conflicting information on long-term trends in bird populations is that future, carefully conducted studies are essential to clarify the details of the trends and identify the factors causing these trends in order to ensure the credibility of bird conservation programs in the eastern deciduous forest (James, McCullough, and Wiedenfeld 1996; Sauer, Pendelton, and Peterjohn 1996).

Forest fragmentation typically is given as the major factor causing long-term declines in the abundance of some Neotropical migratory bird species in the eastern deciduous forest (Lynch and Whigham 1984). Evidence for the role of fragmentation as a major factor affecting declines of these migrants comes from several studies, such as one conducted in the Great Smoky Mountains. The Smokies represents one of the largest and least disturbed forested tracts in the eastern United States; populations of Neotropical migrants did not change significantly in the Smokies from 1947 to 1983 (Wilcove 1988). In contrast, several Neotropical–area-dependent species have become absent or uncommon in many small forested tracts throughout the eastern deciduous forest over roughly this same time period (Whitcomb et al. 1981).

Another logical explanation for a reduction in the abundance of Neotropical migratory birdlife is the loss of wintering habitat to the south in Central and South America. In excess of 50 percent of the migratory songbird species that breed in eastern Canada and the United States overwinter in portions of Central America where more than 25 percent of the wintering habitat will be lost by the end of the century (Diamond 1991). By the early twenty-first century, practically no wintering habitat is expected to remain in Costa Rica except for that in national parks and reserves (Terbough 1989).

In contrast to Neotropical migratory bird species, certain nonmigratory (permanent) bird species in the eastern deciduous forest seem to be affected less by forest fragmentation, such as black-capped chickadees (Lynch and Whigham 1984; Yahner 1985; Maurer and Heywood 1993; Hagan, Vander Haegen, and McKinley 1996). Some short-distance migratory birds, such as eastern bluebirds and Carolina wrens, have exhibited population declines unrelated to fragmentation but instead related to inclement weather conditions, such as the severe winters of 1976–77 and 1977–78 (Sauer, Pendelton, and Peterjohn 1996).

Long-term declines in bird populations are not restricted to forest species. For example, birds associated with grasslands in the northeastern United States also are declining at an alarming rate as these habitats are rapidly being fragmented or lost to forest conversion or agriculture (Vickery 1991). Hence, reclaimed surface mines dominated by grasses, as in northern Pennsylvania, represent valuable habitat for many grassland bird species of concern, such as northern harriers and bobolinks (Yahner and Rohrbaugh 1998).

### *Effects of Forest Size on Select Species*

Individual wildlife species vary considerably in their response to fragmentation of the eastern deciduous forest because of differences in ecological requirements (Whitcomb et al. 1981; Lynch and Whigham 1984). As a general pattern, bird species that nest and feed near forest edges (forest-edge species) are area independent and, hence, are relatively unaffected or perhaps even positively affected by fragmentation. Conversely, species that feed and nest away from a forest edge or depend on large forested tracts (forest-interior species ) (Figure 6.6) show a negative response to fragmentation. For example, in a study of a series of woodlots of various sizes (1.8 to 600 hectares) in Illinois, area-independent species of songbirds (e.g., blue jays) were found both on small, isolated woodlots and on larger, less isolated woodlots, whereas several area-dependent species (e.g., rose-breasted grosbeaks) appeared only in the largest woodlots (Blake 1991).

The size of a forested tract can have a major influence on the occurrence of individual breeding bird species. For instance, the likelihood of encountering at least one breeding pair of an area-dependent species, such as the scarlet tanager, in wooded tracts of less than 10 hectares in

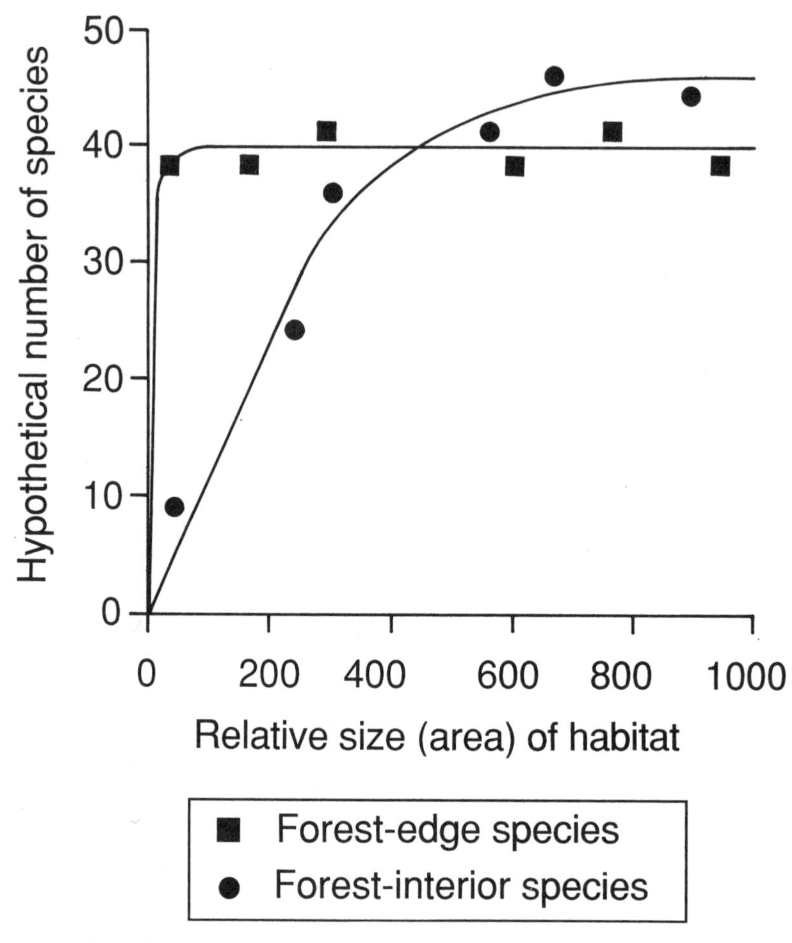


Figure 6.6. A hypothetical species-area curve giving the relative number of forest-edge versus forest-interior species in relation to size (area) of habitat.

Maryland is relatively low (< 50 percent) (Robbins, Dawson, and Dowell 1989) (Figure 6.7). Yet when tracts are greater than 100 hectares, the chance of observing the tanager is much greater (> 70 percent).

Ovenbird biology in relation to forest size has been examined extensively in recent years. In a study of breeding ovenbirds in Pennsylvania, the density of resident males was correlated directly with the size of deciduous woodlots (Porneluzi et al. 1993). Interestingly, and for reasons not clearly understood, a lower percentage of male ovenbirds occupying smaller woodlots (< 70 hectares) were paired with mates than were males in the larger woodlots (> 70 hectares; 47 percent versus 67 percent, respectively). Similarly in Maine, ovenbirds were 37 percent more common in fragments (5 to 60 hectares) than in more contiguous forests (> 500 hectares) but were 40 percent less likely to be paired in fragments compared to those in contiguous forests (Hagan, Vander Haegen, and McKinley 1996). Other studies of ovenbirds, such as those in Missouri (Gibbs and Faaborg 1990) and Quebec (Villard, Martin, and Drummond 1993) also have reported reduced pairing success in ovenbirds as size of forest fragments decreased. In addition, fewer ovenbird nests were found per unit time spent searching in forested fragments than in contiguous forests, which supports the contention that ovenbirds were not reproducing as well in fragmented as in more forested landscapes (Hagan, Vander Haegen, and McKinley 1996). Two factors affecting nesting success in birds in fragmented forested landscapes, nest predation and nest parasitism, will be discussed in detail in the next chapter.

Why are some Neotropical migratory species absent from small forested tracts? Such a forest fragment may simply be too small to accommodate a territory or to provide an adequate supply of food for a breeding pair of birds (Whitcomb et al. 1981). The abundance of insects, which make up the diet of many forest birds, may be reduced in a smaller forested tract (Saunders, Hobbs, and Margules 1991), making these tracts poorer habitat for breeding birds than larger tracts. In some

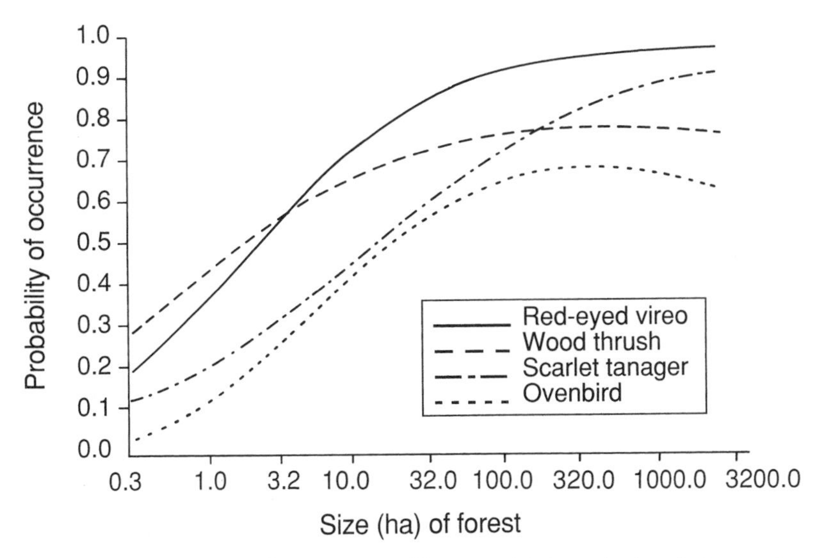


Figure 6.7. Probability of occurrence of four songbird species in relation to size of forest. (Modified from Robbins, Dawson, and Dowell 1989.)

Figure 6.8. A blue-gray gnatcatcher.

cases, as the size of a forested tract becomes smaller, certain features of the habitat that are important to a given species may not exist. For instance, a stream is more likely to be found in a larger woodlot than a smaller woodlot on the basis of chance alone. The Louisiana waterthrush is an example of a bird that does not occur in deciduous woodlots without a source of water. Thus, it is rarely encountered in small woodlots but is more common in larger woodlots, perhaps in response to the greater likelihood of water being present (Robbins, Dawson, and Dowell 1989). Similarly, the presence of fewer blue-gray gnatcatchers (Figure 6.8) in small forested tracts compared to large tracts may be due to a lack of habitat heterogeneity in the smaller tracts. Gnatcatchers nest in bottomland forested habitat; when nestlings are fledged, both parents and young move as a family group to dry upland habitat (Root 1967). As a function of its greater size, a large woodlot generally has a higher chance of containing both bottomland and upland habitats for gnatcatchers than a smaller woodlot.

The effects of woodlot size on bird communities may be affected dramatically by other factors, especially the characteristics of the adjacent landscape (Yahner 1995a). In a study of breeding birds in Ontario woodlots, striking evidence was given to show that both species richness and abundance were markedly lower in 25-hectare woodlots surrounded by houses compared to those in 4-hectare woodlots without nearby houses (Friesen, Eagles, and MacKay 1995). Similarly, in eastern Pennsylvania woodlots, densities of ten breeding bird species increased while eleven species decreased as a possible result of buildings near woodlots (R. G. Mancke and T. A. Gavin, Cornell University, personal communication). These two studies point to the fact that the extent of residential development surrounding woodlots needs to considered when evaluating its impact on urban wildlife.

Large mobile vertebrates that require extensive home ranges to meet their daily requirements often are affected severely by the reduced size of forests. The Florida panther is a good example (Figure 6.9). In presettlement times, the mountain lion ranged throughout North America, and the Florida panther, a subspecies of the mountain lion, likely numbered about fourteen hundred in the Southeast (Simberloff and Cox 1987). Today, probably only about fifty individuals remain in the southern tip of Florida because of recent habitat fragmentation and destruction. The

Figure 6.9. A captive Florida panther. (Photograph by Dennis B. Jordon, U.S. Fish and Wildlife Service.)

panther's future still looks bleak despite the establishment of the 12,140-hectare Florida Panther National Wildlife Refuge in southern Florida by the U.S. Department of the Interior in 1989 (Matthews 1991).

The black bear is another mobile, wide-ranging vertebrate that is influenced by habitat size. Although the black bear is not a threatened species in North America, some isolated populations in the southeastern United States are in jeopardy (Pelton 1982). Forest fragmentation and loss, bear-vehicular collisions, and poaching have taken their toll on these southeastern bear populations. As a result, sanctuaries have been established in some states, e.g., North Carolina, over the past few decades as a viable means of providing suitable habitat for bears and reducing incidences of poaching (Powell et al. 1996). Law-enforcement programs also have helped to curb the illegal killing of black bears for gall bladders and other body parts throughout the eastern deciduous forest. One notable program was the well-publicized Operation Smoky, which was conducted by personnel from the National Park Service, U.S. Fish and Wildlife Service, Tennessee, and North Carolina in the late 1980s and resulted in the arrest and conviction of several poachers of black bears.

Fragmentation of the eastern deciduous forest can affect the behavior of wildlife, but this phenomenon has not been given adequate attention by wildlife biologists (Yahner and Mahan 1997a). For instance, wildlife biologists have inventoried and monitored breeding bird populations for many decades in a variety of fragmented and unfragmented landscapes using standardized counts of singing males (e.g., Robbins 1970). Several recent studies, however, have shown that song rates during the breeding season may depend on whether a male is paired or the extent of forest fragmentation. Rates of singing in unmated ovenbirds and Kentucky warblers were 3.5 and 5.4 times higher, respectively, than those of mated males (Gibbs and Wenny 1993). In another study, both ovenbirds and wood thrushes sang significantly more often in contiguous forests (> 50 hectares) than in forested fragments (5 to 24 hectares); conversely, northern cardinals sang more often in the forested fragments (McShea and Rappole 1997). Thus, wildlife biologists conducting long-term monitoring programs of breeding birds in fragmented landscapes must remember that certain species may vocalize at different rates and, hence, be detected to different degrees.

Forest fragmentation also may influence the behavior of mammals. Eastern chipmunks, for example, spend more time foraging in habitats fragmented by clear-cutting compared to chipmunks in more contiguous forest, perhaps as a result of lower food resources in a fragmented forest (Mahan 1996). Home-site selection by chipmunks, however, did not differ in chipmunks occupying fragmented versus contiguous forest stands (Mahan and Yahner 1996). Another forest mammal, the red squirrel, is highly territorial in boreal forests (Smith 1968) but not in isolated woodlands (e.g., farmstead shelterbelts) in the Midwest (Yahner 1980). Herd or group size in the white-tailed deer is related to habitat openness; in more fragmented habitats, the number of deer in a group is higher than in less fragmented habitats (Hirth 1977). Perhaps not surprisingly, an Illinois study showed that deer were more vulnerable to hunting in highly fragmented landscapes with small amounts of cover than in landscapes containing more cover and less fragmentation (Foster, Roseberry, and Woolf 1997).

In summary, an understanding of how forest fragmentation affects the abundance and the distribution of wildlife populations and communities is very important to conservation. But equally relevant to understand is the behavioral response of individual animals to the future consequences of fragmentation in the eastern deciduous forest (Lima and Zollner 1996; Yahner and Mahan 1997a).

### *Effects of Forest Isolation on Select Species*

Increased isolation of forested tracts caused by fragmentation can have a profound effect on movements of some wildlife species. For example, the southern flying squirrel (Figure 6.10) is found throughout much of the mature deciduous forest in the eastern United States. This squirrel relies on gliding to move efficiently from tree to tree and to escape detection by predators, but its movements are impeded by large forest clearings (> 75 meters in width) (Bendel and Gates 1987).

In intensively farmed landscapes, the absence of woody or shrubby corridors (narrow strips of land that differ from the adjacent habitat on each side [Forman and Godron 1986]) connecting woodlots can restrict movements of eastern chipmunks and eastern cottontails between woodlots (Wegner and Merriam 1979; Swihart and Yahner 1982). Even with birds, which typically are more mobile than most mammals, the degree of isolation of suitable habitat can influence the distribution of populations (Dunning et al. 1995; see chapter 7). Thus, the effects of fragmentation depend on the extent of isolation created by fragmentation as well as on the normal movement patterns and dispersal capabilities of the species in question (Harris and Silva-Lopez 1992). Furthermore, when

Figure 6.10. A southern flying squirrel. (Photograph by J. Timothy Kimmel, School of Forest Resources, Pennsylvania State University.)

an extensive network of roadways is superimposed upon a fragmented habitat, the potential for incidences of wildlife-vehicular collisions increases, as with deer populations in and around Gettysburg National Military Park, Pennsylvania (Storm, Cottam, and Yahner 1995). In Florida, such a network is the major cause of death for large mammals, such as Florida panthers and black bears (Harris and Gallagher 1989). Nearly one-half (49 percent) of the deaths in Florida panthers are attributed to vehicular collisions (Maehr, Land, and Roelke 1991), giving rise to the construction of underpasses for wildlife along highways with heavy vehicular traffic in Florida (Foster and Humphrey 1995). These underpasses have been monitored and shown to be used regularly by a variety of wildlife, including Florida panthers, white-tailed deer, bobcats, raccoons, and alligators.

## Forest Fragmentation and Fish Habitat

The impact of forest fragmentation on fishes and other aquatic organisms, such as invertebrates, is becoming an important concern to natural resource managers in the eastern United States and throughout many areas of the world (Moyle and Leidy 1992). Removal of forest cover near small headwater streams can reduce the amount of organic material entering the stream, thereby affecting food abundance and nutrient input for aquatic organisms (Hesser et al. 1975). Moreover, as discussed in the last chapter, the cutting of trees adjacent to streams can increase water temperature and sediment and decrease woody debris in waters, soon making the habitat unsuitable for trout and valuable fishes in streams of Pennsylvania and other northeastern states (Hesser et al. 1975; Guildin 1990; Gullion 1990). Public lands, like national forests, contain valuable and extensive lakes, reservoirs, and fishable streams (Hollingsworth 1988). Hence, preservation and careful manipulation of vegetation in aquatic environments on public lands can go a long way toward maintaining or restoring fish populations in habitats degraded by forest fragmentation.

## Forest Fragmentation and Changes in Microclimate

We have seen that fragmentation of the eastern deciduous forest affects the size and the isolation of tracts and, therefore, the biota associated with these tracts. Yet in a more subtle way, fragmentation can also alter the microclimate of the tracts, causing direct or indirect effects on forest plants and animals. These changes in the microclimate of forested tracts may be measured in terms of changes in radiation fluxes, wind, and water (Saunders, Hobbs, and Margules 1991).

### *Radiation Fluxes and Fragmentation*

A forested tract that is clear-cut or cleared for crop production receives a greater amount of solar radiation (energy) at ground level than an undisturbed forested tract. As a consequence, daytime temperatures are generally higher in cleared tracts than in undisturbed forested tracts, although this may differ somewhat with season or crop type (Geiger 1965). In addition, increased radiation can reduce the amount of soil moisture, especially along the edges of tracts (Ranney, Bruner, and Levenson 1981).

Because of these differences in solar radiation on cleared versus forested tracts, the structure and composition of plant communities may vary between edges and interiors of forested tracts (Ranney, Bruner, and Levenson 1981) (Figure 6.11). Shade-tolerant species of plants in the eastern deciduous forest, such as beech, may occur only in the interiors of forested tracts, whereas shade-intolerant species, such as ash, may be restricted primarily to the edges of the tracts (Saunders, Hobbs, and Margules 1991). Newly created edges generally are dominated by a dense understory of herbaceous perennials and woody seedlings

Figure 6.11. Edge created by forest clear-cutting. (Photograph by author.)

or saplings because of increased solar radiation (Ranney, Bruner, and Levenson 1981). These differences in plant structure and composition can influence the distribution and abundance of wildlife associated with edges of forested tracts, such as songbirds that require a certain density or type of understory vegetation for foraging and nesting (Bond 1957; Lynch and Whigham 1984; Yahner 1987).

Forest fragmentation can encourage the invasion of exotic plants, many of which are pioneer or early-successional species, into the remaining tracts because of microclimatic changes and increased ratios of forest to nonforest land (Brothers and Spingarn 1992). In Indiana, for instance, the number of exotic plant species was relatively high along the edges of forested tracts adjacent to agricultural fields. An average of eleven exotic species invaded the initial 2 meters of the field-forest boundary, compared to less than two species 8 meters or more from the boundary; dandelion, multiflora rose, and goosefoot were among the most common exotic species occurring near the forest edge. Furthermore, exotic species were more diverse and frequent along warm (south and west) edges than along cool (north and east) edges, presumably because of microclimatic differences.

Road and trail corridors in Glacier National Park, Montana, have enabled exotic plant species to invade grassland habitats, which could have negative consequences for native flora and fauna (Tyser and Worley 1992). Similarly, the fauna of national parks in the eastern deciduous forest, such as Gettysburg National Military Park and Valley Forge National Historical Park, Pennsylvania, consists of a high proportion of exotic plant species; 22 percent of the total plant species are exotic at Gettysburg (Yahner et al. 1992), and 34 percent are exotic at Valley Forge (Cypher et al. 1986). Because the current policy of the National Park Service is to manage only for native species (National Park Service 1988), the invasion of these public lands by exotics is of much concern (Houston and Schreiner 1995). Some scientists, for instance, view the invasion of exotic plants on natural or managed public and private lands as a major threat to biodiversity because they compete and hybridize with native species or otherwise affect natural ecosystem processes (Hobbs and Humphries 1995; Levin, Francisco-Ortega, and Jansen 1996).

### *Wind and Fragmentation*

Removal of tree cover by forest fragmentation can affect airflow over the landscape (Saunders, Hobbs, and Margules 1991). Changes in airflow can result in a greater likelihood of damage to forest vegetation in the remaining forested tracts, either physically through windfall or through increased evapotranspiration (Ranney, Bruner, and Levenson 1981). Increased wind exposure also can remove loose bark from overstory trees that serves as substrate for bark-dwelling invertebrates (Saunders, Hobbs, and Margules 1991). This change in turn can affect the availability of food resources for forest birds, such as downy woodpeckers and black-capped chickadees, that forage extensively on rough-barked trees in search of insects and other invertebrates (Brawn, Elder, and Evan 1982; Rollfinke and Yahner 1991). Moreover, the foraging behavior of some bird species in the eastern deciduous forest may be influenced by increased wind along edges of forested tracts; for instance, the downy woodpecker tends to prefer the interior rather than the edge of forested tracts during winter, perhaps because the interior has a more favorable microclimate for birds while foraging (Yahner 1987).

### *Water Flux and Fragmentation*

Forest fragmentation, such as occurs when a forested tract is replaced with agricultural crops or pasture, can modify the local hydrological cycle (Saunders, Hobbs, and Margules 1991). Removal of forest cover reduces the interception of rainfall, evapotranspiration, and surface and soil moisture (Kapos 1989). These changes then can lead to differences in the decomposition of leaf litter and in the amount of ground-level vegetation (Saunders, Hobbs, and Margules 1991), which can have obvious and profound effects on the distribution and abundance of forest birds, such as wood thrushes and ovenbirds, and small mammals, such as southern red-backed voles and masked shrews, that forage on the forest floor (Mastrota, Yahner, and Storm 1989; Yahner 1989; Rollfinke, Yahner, and Wakeley 1990). Red-backed voles and masked shrews are dependent especially on the availability of moist microenvironments within suitable forested habitats (Getz 1961; Powell and Brooks 1981).

## Forest Fragmentation—How Big Is Enough?

How big is enough? has been asked by wildlife biologists at least since the 1930s (Shafer 1995). It will remain a persistent question for scientists, natural resource managers, and society as forest fragmentation continues to loom as a contemporary conservation issue. Let us first briefly consider the current plight of three wide-ranging carnivores, the grizzly (or brown) bear, the black bear, and the gray (or timber) wolf. The grizzly bear in the lower forty-eight states occurs principally in the Greater Yellowstone Area (GYA includes Yellowstone and Grand Teton National parks, plus five adjoining national forests), which is home to only a few hundred of the remaining grizzlies. Although the GYA is nearly 23,000 square kilometers in size, it is theoretically big enough to provide space for the home ranges of only six adult male grizzlies—in short, the GYA may be "too little, too late" to save the species. Because of habitat fragmentation and reduced access to habitats free of human disturbances in the western states, the outlook for grizzly populations in the continental United States is bleak (Mattson and Reid 1991).

Similarly, in the deciduous forest of the Southeast, habitat may be too fragmented and of inadequate size to sustain viable populations of black bears in the future (Burdick et al. 1989; Harris and Silva-Lopez 1992). As mentioned earlier, sanctuaries with suitable habitat devoid of hunting and poaching are vital to the long-term persistence of black bear populations in the Southeast (Powell et al. 1996). Each sanctuary should probably be at least 400 square kilometers in size, with a low road density to minimize bear-vehicular collisions and poachers' access to the sanctuary.

Another wide-ranging carnivore, the eastern gray wolf, occurs in southeastern Canada but has been extirpated from the northeastern United States; the southeastern Canada wolf is a separate subspecies from that of the upper Midwest states (Nowak 1995). A contiguous tract of suitable wolf habitat from northeastern Vermont to Maine, however, has been identified in the event that restoration of the eastern gray wolf into the Northeast becomes a reality in the near future (Mladenoff and Sickley 1998). This area is approximately 53,000 square kilometers in extent and is capable of supporting more than one thousand wolves, which exceeds the minimum of 25,000 square kilometers recommended by the U.S. Fish and Wildlife Service for the successful recovery of gray wolf populations.

How big is enough? is not a question limited to terrestrial wildlife. Restoration of the Oklawaha River as well as an expansion of the Ocala National Forest in Florida would provide an extensive freshwater habitat for manatees, which are large, migratory mammals (Harris and Silva-Lopez 1992). Restoration and expansion of aquatic habitat in Florida would make the Ocala National Forest the first national forest to serve specifically as habitat for the conservation of an endangered marine mammal.

This question How big is enough? is pertinent to ecosystem management. In the Pacific Northwest, the preservation of the ancient forest ecosystem and the regional economics of the timber industry are important issues (Hill 1992). In the early 1990s, the U.S. Department of the Interior proposed the establishment of a 3.6 million-hectare preserve in the Pacific Northwest to save the old-growth forest, but the U.S. Department of Agriculture proposed only a 2.8 million-hectare preserve. Are 3.6 million or 2.8 million hectares adequate to conserve this important ecosystem? The difference between the proposed preserve sizes is equivalent to about 2.5 times the size of Rhode Island—and is a difference of about 0.8 billion board feet in marketable lumber. Moreover, is the 23,000 square kilometer GYA of an adequate size to conserve its ecosystem? As we have seen, it is too small to sustain a grizzly population. However, the GYA may be a minimum dynamic area (MDA) (Pickett and Thompson 1978) in terms of its ability to withstand disturbances, such as those created by the extensive fires in the late 1980s. The concept of an MDA is clearly relevant to our discussion of both How big is enough? and ecosystem management in the eastern deciduous forest, because an MDA is intended to be the smallest area needed to maintain natural disturbances in the ecosystem. An MDA also is expected to serve as an internal source of colonization by species, thereby reducing the risk of localized population extinctions.

In addition, the question How big is enough? is relevant to forest management. Browsing by white-tailed deer on seedlings can prevent the regeneration of desirable timber species in Wisconsin, such as eastern hemlock and white cedar (Alverson, Waller, and Solheim 1988). In Wisconsin, setting aside diversity maintenance areas of 200 to 400 square kilometers in mature, contiguous forests has been proposed to mitigate deer damage to regenerating forests. These areas would be kept free of grassy openings and clear-cut stands, making them less suitable habitat for deer and perhaps ensuring the regeneration of timber species and the preservation of other plant life. Suits were filed against the U.S. Forest Service to ensure the establishment of diversity maintenance areas in two national forests (Nicolet and Chequamegon) in northern Wisconsin (Mlot 1992).

These examples and others discussed earlier dealing with forest size and bird diversity (e.g., Galli, Leck, and Forman 1976) point out the need to ask the question How big is enough? and to send a profound conservation message to natural resource managers and land-use planners—that "bigger is better" in many circumstances. When designing nature reserves or acquiring lands for the benefit of wildlife, a conservation goal in the eastern deciduous forest should be to create or preserve the largest forested tracts when possible. Unfortunately, natural resource managers must deal with what forest remnants remain after decades of fragmentation, and they have little opportunity to actually "create" large forested reserves (Saunders, Hobbs, and Margules 1991). This situation reaffirms the extreme importance of maintaining large, existing tracts of public lands, such as state and federal parks, for the long-term perpetuation of ecosystems and their associated wildlife species in the eastern deciduous forest (Wilcove, McLellan, and Dobson 1986).

On the other hand, the value of small forested tracts should not be based solely on the number of species they contain (Shafer 1995). A small tract may be very important in educating the general public about the natural world or may be used as a monitoring site, such as in a more urbanized landscape. As we shall see in the next chapter, remnant forested tracts can be made more beneficial to biota if natural resource managers make every effort to provide connectivity with other forested tracts in the landscape. Thus, natural resource managers need to view the bigger picture from a landscape perspective while simultaneously focusing on the value of individual tracts to ensure the long-term conservation of wildlife in the eastern deciduous forest (e.g., see Petit, Petit, and Martin 1995).

# 7. CORRIDORS AND EDGES IN RELATION TO FRAGMENTED FORESTS

## Corridors and Forest Fragmentation

As we have seen in earlier chapters, the original forest of the eastern United States consisted of expansive, unbroken tracts, except for occasional clearings created by natural events or Native Americans. Today, the eastern deciduous forest is relatively discontinuous, with forested tracts of various sizes amid an agricultural or urban landscape. Many of these forested tracts are isolated from others, thereby resembling a series of islands in the landscape—a situation quite unlike that present before European settlement. Wooded corridors in the landscape, however, counter the effects of forest fragmentation by connecting these isolated tracts, thereby better approximating conditions found in a pristine forest. Linear habitats, such as fencerows, shelterbelts, and strips of wooded habitat, help provide connectivity of landscapes in the eastern deciduous forest. Furthermore, because corridors add aesthetic value to the landscape, the idea of establishing or protecting existing corridors, such as "greenbelts" or "landscape linkages," has much popular appeal in land-use planning and wildlife conservation strategies (Noss 1987; Harris and Gallagher 1989; Lacasse 1994).

Just how valuable to wildlife are corridors? How does the quality (e.g., width, vegetative composition) of corridors affect their use by wildlife? Although definitive answers to these questions are lacking (Harrison 1992; Demers et al. 1995), corridors have some positive wildlife values and important conservation implications to the ecology of the eastern deciduous forest (e.g., Noss 1987; Simberloff and Cox

1987; Rosenberg, Noon, and Meslow 1997). One positive value of corridors is that they facilitate and perhaps enhance movements by wildlife among forested tracts. Without corridors, some forest wildlife species would be unable to disperse or move to and from isolated woodlots because of nonforested barriers, such as agricultural croplands (Noss 1983; Harris and Silva-Lopez 1992). In particular, small forest mammals with limited dispersal ability, such as eastern cottontails and white-footed mice, may benefit by the presence of wooded corridors in an agricultural landscape (Swihart and Yahner 1982; Yahner 1983a; Henein and Merriam 1990). Although birds are more capable of moving between isolated woodlots than small mammals are, movements by birds between woodlots in agricultural areas occur more readily if corridors are present in the landscape. For instance, adult American robins and brown thrashers moved more often between patches of wooded habitat connected by wooded corridors than between isolated wooded habitats (Haas 1995). Similarly, movements by Bachman's sparrows between clear-cut stands were enhanced when connected by linear corridors of early-successional habitat (Dunning et al. 1995).

Little is known about what constitutes optimum corridors for wildlife. For small mammals, corridors several meters wide and containing shrubby vegetation probably have more value than narrower, herbaceous fencerows in extensively farmed landscapes (Merriam and Lanoue 1990). In forested landscapes, birds readily chose 100-meter-wide wooded buffer strips as dispersal and movement corridors over nearby clear-cut stands (Machtans, Villard, and Hannon 1996). If corridors are to serve as an effective means of allowing large animals to move regularly among forested tracts in a fragmented landscape, the corridors probably should be at least as wide as the typical diameter of a home range (Harrison 1992). As an example, a corridor for female white-tailed deer should be at least 2 kilometers wide, which is approximately the diameter of the home range of an adult female. The optimal width of a suitable corridor could be determined by several other factors, such as its length, the topography, and the biology (e.g., food habits) of the target species (Beier and Loe 1992; Lindemayer and Nix 1993). Herbivores, for instance, may require a different corridor design than carnivores or omnivores. In forested landscapes, open corridors, e.g., fields, are probably important dispersal routes for aesthetic invertebrates, such as butterflies (Sutcliffe and Thomas 1996). In some cases, stepping stones of suitable habitat may be more valuable to dispersing butterflies than a linear corridor of habitat (Schultz 1998).

A second benefit of corridors, assuming they are of sufficient width, is that they provide valuable habitat for wildlife. Wooded riparian corridors at least 100 meters in width, for example, are important habitat for breeding forest birds, such as red-eyed vireos and white-eyed vireos (Hodges and Krementz 1996) (Figure 7.1). Similarly, uncut, 100-meter-wide strips of forest are much more important as habitat than small, uncut plots (1 hectare) to forest-canopy breeding birds (e.g., red-eyed vireos) or to insectivorous wintering birds (e.g., woodpeckers) (Yahner 1997a). In some bottomland hardwood forests, the conservation of diverse breeding bird communities, however, may require corridors at least 500 meters in width (Kilgo et al. 1998). Open corridors created by transmission rights-of-way in forested landscapes are valuable to breeding birds that require early-successional habitats; in Pennsylvania and

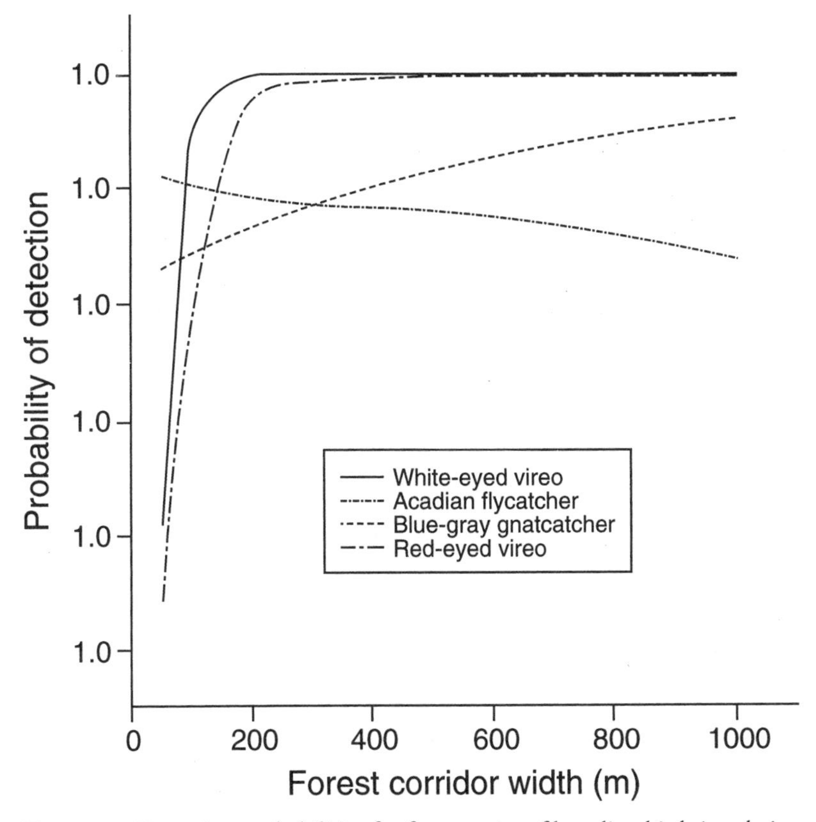


Figure 7.1. Detection probabilities for four species of breeding birds in relation to width of wooded riparian corridors along the Altamaha River, Georgia. (Modified from Hodges and Krementz 1996.)

elsewhere in the Northeast, rights-of-way that are 60 to 70 meters wide contain a diverse breeding bird community (Bramble, Yahner, and Byrnes 1992; Askins 1994; Yahner 1995a).

The value of corridors to wildlife as either movement corridors or habitat may also depend on land uses or human activities adjacent to the corridors (Beier and Loe 1992). For instance, based on studies of salamanders, the quality of a given wooded corridor is greater if it contains habitat more similar to the forested tracts connected to it compared to habitat adjacent to the corridor in the surrounding matrix (Rosenberg, Noon, and Meslow 1997). Species sensitive to forest size, such as acadian flycatchers and red-eyed vireos, were among five bird species more common in wooded riparian corridors adjacent to farmlands than in corridors adjacent to urban areas in Florida (Smith and Schaefer 1992). Conversely, edge species, such as American crows and blue jays, occurred more often in urban corridors. In the western United States, juvenile mountain lions tended to avoid wooded corridors that contained human dwellings (Beier 1995).

A third value of corridors is that they may allow the exchange of individuals among populations and reduce the likelihood of populations becoming extirpated. Using the mountain lion again as an example, isolated populations in the West have a better chance of persisting when corridors connect suitable habitat for these cats (Beier 1993). Thus, because corridors provide connectivity among a landscape's populations, which is often termed a metapopulation (Figure 7.2) (Noss 1987; Bennett 1990; Henein and Merriam 1990), a higher regional population size is better ensured. A larger population size is more advantageous than a smaller population size because genetic variation tends to increase as the size of a population increases, thereby mitigating the negative impacts of genetic drift or inbreeding on the population (Harris 1984; Noss 1987).

A metapopulation comprises spatially distinct populations, some of which may go extinct locally but be recolonized later by individuals dispersing among populations (Wells and Richmond 1995). A good example of metapopulations are those of the New England cottontail, which is a candidate for federal threatened or endangered status (Litvaitis and Villafuerte 1996). Small numbers of this cottontail occur in disturbed sites, such as young even-aged stands, within a forest-dominated landscape. In order to ensure the long-term survival of a New England cottontail metapopulation, a conservation recommendation is to establish a network of early-successional habitats with periodic disturbances

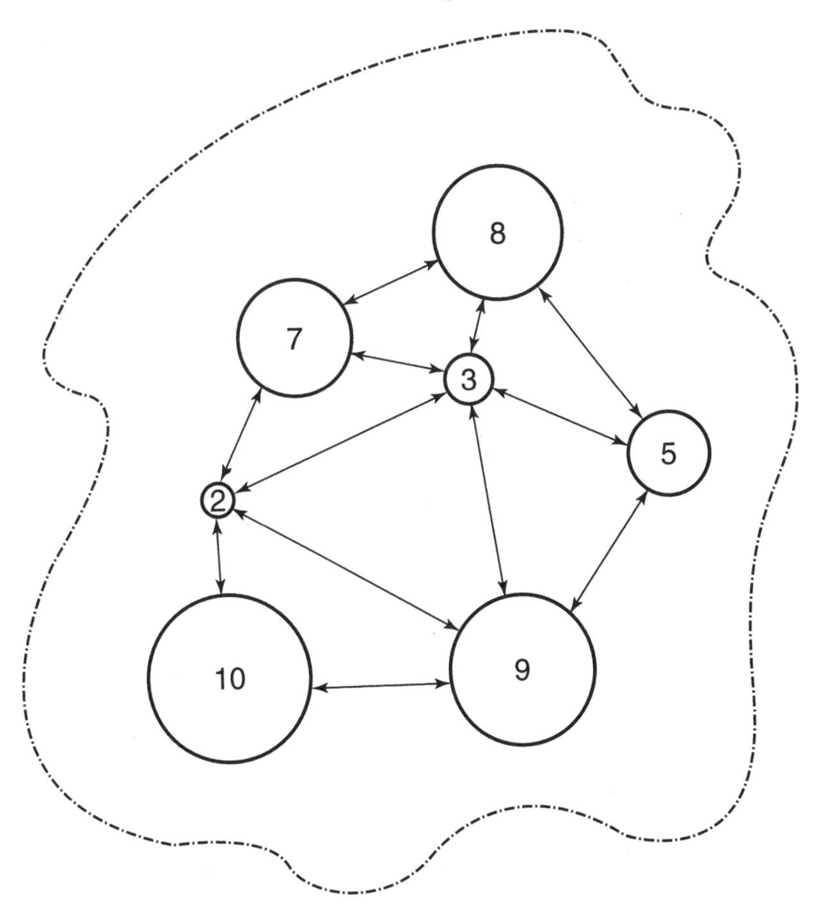


Figure 7.2. A hypothetical landscape containing a group of interconnected populations, termed a metapopulation. Populations are shown by circles, and the relative size of an individual population is indicated by a number contained within a given circle. (Modified from Litvaitis and Villafuerte 1996.)

(e.g., cutting, burning). However, to be of value to a cottontail metapopulation, each habitat needs to be relatively large (15–75 hectares) and relatively close (< 1 kilometer) to other disturbances, prerequisites based on the known dispersal distance of individual cottontails. Another wildlife species that has been given considerable attention from a metapopulation perspective is the threatened Appalachian woodrat in Pennsylvania (Balcom and Yahner 1996). Small populations of woodrats occupy rocky outcrops throughout forested, mountainous areas of Pennsylvania. Over the past few decades, however, woodrat populations have been extirpated from sites known to be occupied in the past, while other populations have been documented from sites not known to be used by woodrats in recent decades.

The creation of a mega-corridor or landscape linkage at the border of Florida and Georgia demonstrates nicely how natural resource managers can provide connectivity of large forested tracts on a regional scale (Figure 7.3). The landscape linkage between Okefenokee Swamp and Osceola National Forest via Pinhook Swamp and similar linkages will likely play a major role in ensuring the survival of the remaining Florida panthers, black bears, red-cockaded woodpeckers, and other wide-ranging wildlife in the Southeast (Harris and Gallagher 1989; Harris and Silva-Lopez 1992). Landscape linkages are also important as corridors to many bird species that migrate long distances. These corridors provide potential stopover or staging sites where birds, such as migratory warblers and American woodcock, can periodically rest, feed, and re-

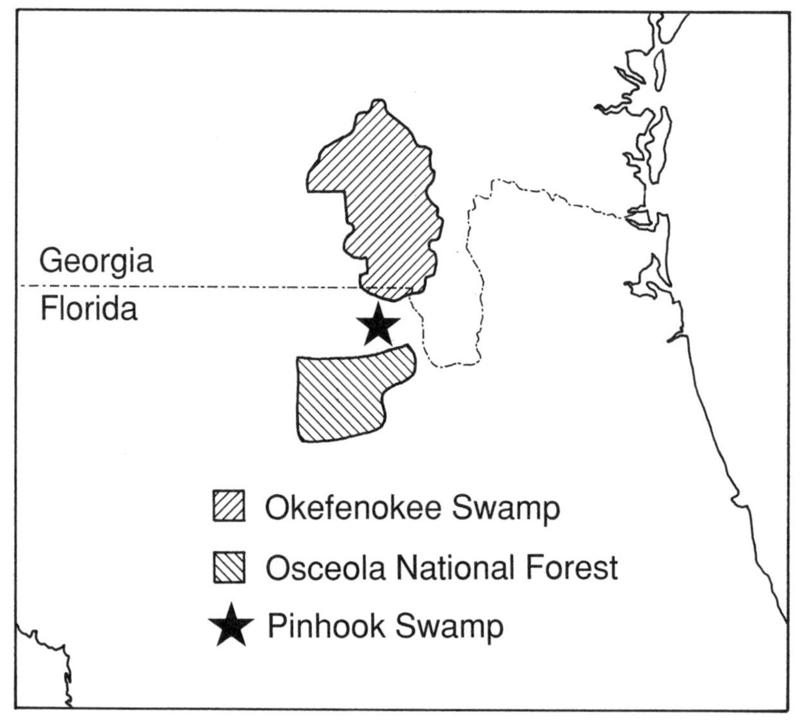


Figure 7.3. Creation of a landscape linkage in Florida and Georgia. Here the Okefenokee Swamp is "linked" to the Osceola National Forest by the Pinhook Swamp. (Modified from Harris and Gallagher 1989.)

plenish energy reserves (Martin 1980; Moore and Yong 1991; Krementz, Seginak, and Pendleton 1994; Morris, Richmond, and Holmes 1994).

In the Allegheny National Forest of northwestern Pennsylvania, wooded corridors totaling 33,200 hectares have been identified as large-scale linkages to connect and manage eight isolated old-growth stands (Nelson et al. 1997). The size of these corridors ranges from 526 to 4,050 hectares, and the width varies from about 400 to 800 meters. Thus, these examples of grand-scale landscape linkages in Florida and Pennsylvania show that they may have different but extremely important primary functions; hence, efforts to maintain or establish landscape linkages in the eastern deciduous forest should continue.

In cases where landscape linkages have not or cannot be established, attempts should be made to acquire lands in proximity to one another to better ensure the preservation of wildlife diversity or a species of special concern. For instance, numerous state and federal management areas have been set aside in Michigan for the conservation of the endangered Kirtland's warbler (Figure 7.4). The acquisition of these management areas in a localized area, combined with appropriate habitat-management practices (e.g., planting of jack-pine plantations), will better safeguard against the extirpation of Kirtland's warbler metapopulations (Probst and Weinrich 1993).

Alternatively, corridors may negatively influence some wildlife (Noss 1987; Simberloff and Cox 1987; Hess 1994). Corridors may facilitate the spread of disease, noxious species, or fire, which may be detrimental to desirable species of wildlife. As mentioned in chapter 6, logging roads and other linear corridors may facilitate invasion by exotic plants, thereby having a detrimental effect on native plants (Levin, Francisco-Ortega, and Jansen 1996). In Australia, the cane toad was introduced in 1935, causing a serious negative impact on the native fauna (Seabrook and Dettmann 1996). This amphibian is poisonous to predators and has outcompeted other species for food, shelter, and breeding space by reaching densities of three thousand toads per hectare! Simulation models have provided some evidence that wildlife diseases, e.g., the highly contagious disease called canine distemper that affects carnivores, are capable of spreading rapidly within a metapopulation via corridors (Hess 1994). In recent decades, roads built for oil exploitation have been used by red fox to colonize northern Alaska; as a result, the red fox has outcompeted the native arctic fox for important denning sites (Rudzinski et al. 1982). Finally, corridors may be viewed as negative because their maintenance or acquisition can be costly. Yet what price is too much

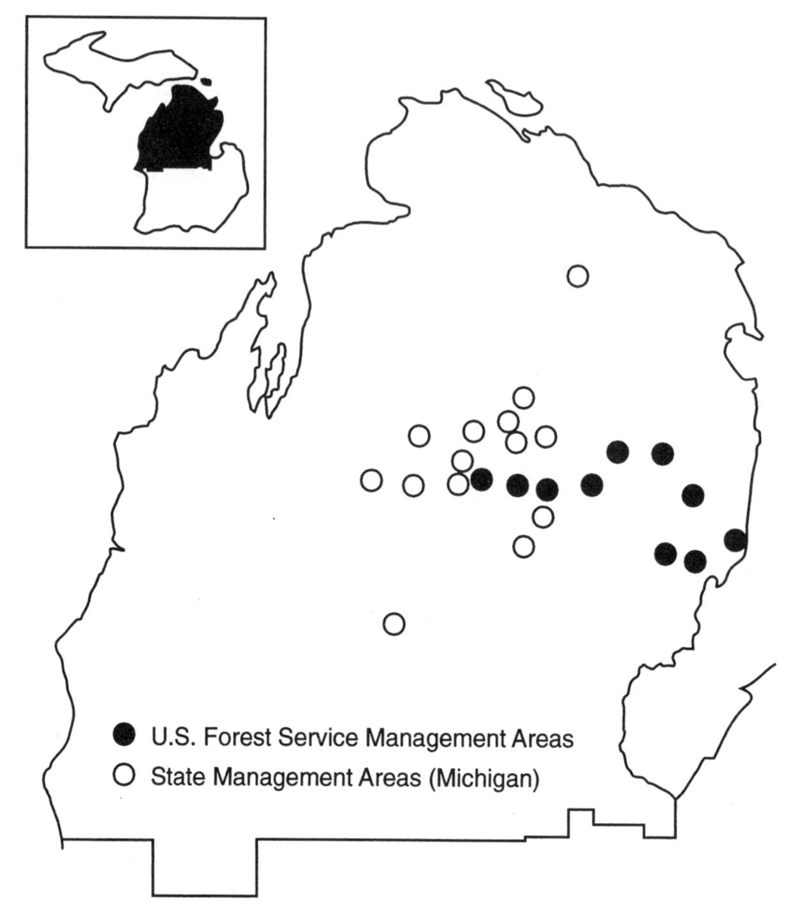


Figure 7.4. Distribution of state and federal management areas for Kirtland's warblers, an endangered species, in Michigan. (Modified from Probst and Weinrich 1993.)

when maintenance of biodiversity of forest wildlife and species of "special concern" are at stake?

Corridors will not be the only answer to problems associated with the conservation of biodiversity and declining species of wildlife in the eastern deciduous forest (Noss 1987). However, the advantages of corridors to the landscape seem always to outweigh the disadvantages. Future management of the eastern deciduous forest and its wildlife must transcend earlier notions that the job of natural resource managers is complete after a park, a refuge, or a national forest is established. The next necessary step is to ensure that natural ecosystems, including all species of forest biota on these lands, can function with minimal disturbance by human activities; a means to achieve this goal is to provide greater connectivity in the eastern deciduous forest (Harris 1984; Harris and Gallagher 1989; Harris and Silva-Lopez 1992).

## Edges and Forest Fragmentation

An edge (or ecotone) may be defined as the interface between two distinct landscape elements (e.g., plant species of different structure, plant communities, successional stages of a forest stand, or land uses) (Thomas, Maser, and Rodiek 1979; Forman and Godron 1986). Edges are either inherent or induced. An inherent edge is a long-term feature of a forested landscape that often is a function of differences in soil type, topography, or microclimate (Thomas, Maser, and Rodiek 1979). For example, an inherent edge exists at the interface of a deciduous forest stand and a coniferous stand. In contrast to an inherent edge, an induced edge is usually short-lived and can be the result of human activities or natural events, such as wildfires or windstorms. An induced edge occurs at the boundary of a mature forest with either a clear-cut stand or an agricultural field (Yahner 1988a).

Because induced edges are often a direct result of forest fragmentation, they are of particular interest to natural resource managers concerned with the impact of fragmentation on forest biota. As forest fragmentation increases, the amount of edge typically increases (Figure 7.5).

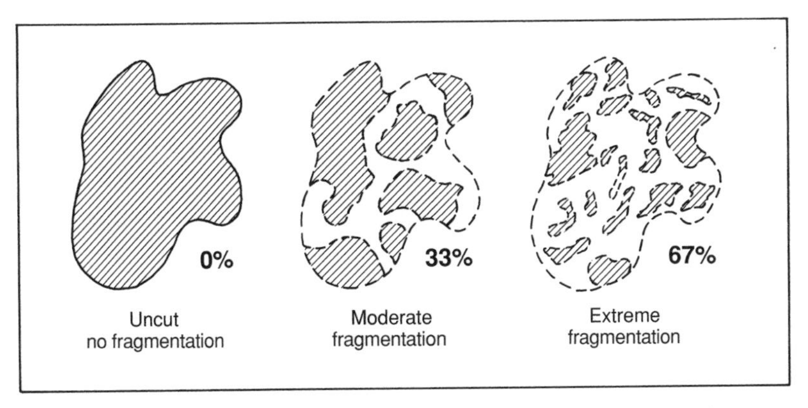


Figure 7.5. The amount of edge can increase with greater amounts of forest fragmentation. In the uncut tract (0 percent fragmentation), no edge is present; in the tract with extreme fragmentation (67 percent), considerable edge is present.

Furthermore, as the size of forested tracts becomes smaller or less circular, the proportion of edge to interior habitat becomes greater (Figure 7.6) (Forman and Godron 1986).

## History of the Edge Concept

Since the publication of the classic textbook *Game Management* (Leopold 1933), edges have been viewed by forest and wildlife managers as beneficial to wildlife (Ratti and Reese 1988; Yahner 1988a). Leopold noted that wildlife communities were often more diverse along forest edges than along forest interiors. Leopold concluded that greater wildlife diversity along edges, often termed an edge effect, was due in part to either an increased variety of vegetation along edges compared to areas away from edges or to the existence of two different habitats—or the "best of two worlds"—in proximity. For instance, habitat for ruffed grouse in central Pennsylvania is being managed using a checkerboard pattern of small (1 hectare) clear-cut plots of different ages (Yahner 1984) (Figure 6.5). Grouse habitat management has resulted in a more diverse community of breeding birds than that found in a nearby uncut forest stand, in part due to the creation of increased amounts of edge habitat (Yahner 1984, 1993a).

A second reason why edges have been viewed as being very favorable to forest wildlife is that until the 1960s, the wildlife profession managed almost exclusively for "game" or consumptive species (Zagata 1978;

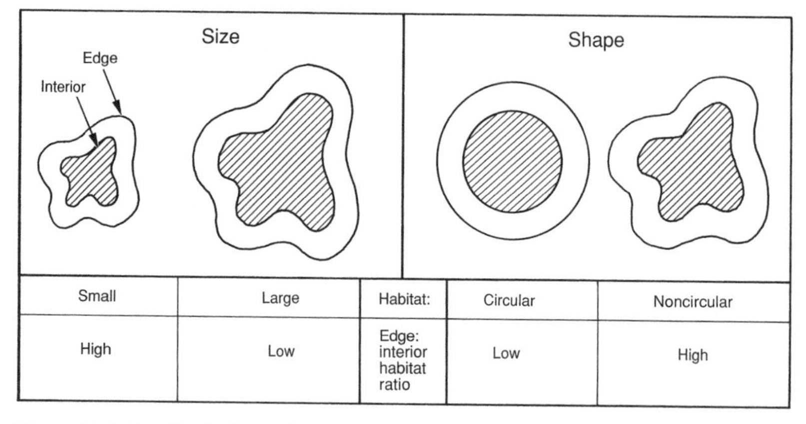


Figure 7.6. Smaller habitats have a higher edge to interior habitat ratio than larger habitats, and noncircular habitats have a higher edge to interior habitat ratio than circular habitats.

Bolen and Robinson 1995). White-tailed deer and eastern cottontails, for instance, have always been two of the most important and intensively managed game species in the eastern deciduous forest (Chapman, Hockman, and Edwards 1982; Hesselton and Hesselton 1982). Both are considered "edge" species, and management recommendations for deer and cottontail habitat include the creation of edge or brushy habitat conditions (Burger 1973).

## Wildlife Species Benefited by Edge and Fragmentation

An edge species may be defined as one that conducts most or all of its daily activities at or near edges (Forman and Godron 1986). But this rather simplistic definition is difficult to apply to various species for at least three reasons (Yahner 1988a). First, edges may be used for a variety of purposes, depending on the species. In farmlands, mourning doves may be considered an edge species because they nest along edges of small woodlots or shelterbelts that consist largely of edge habitat; however, doves may move considerable distances away from an edge into nearby agricultural or old fields to feed (Yahner 1982, 1983b). Second, some species may be defined as edge species in one season but not in another. Eastern cottontails in intensively farmed habitats may rely on small woodlots or shelterbelts in autumn after farm crops are harvested and in winter when herbaceous cover is minimal; however, in summer, woodlots or shelterbelts may seldom be used (Swihart and Yahner 1982). Third, whether a species can be classified as an edge species may vary geographically. The red squirrel is typically found in extensive coniferous forest in northern latitudes; in contrast, this red squirrel may occupy small woodlots or shelterbelts in more southerly latitudes (Yahner 1980; Mahan and Yahner 1992). Although we may encounter some difficulty in defining an edge species, certain species of wildlife have benefited from increased edge and forest fragmentation in the eastern United States.

### *Some Representative Mammals Benefited by Edge and Fragmentation*

The white-tailed deer is a classic example of a species that has increased in numbers over the past few decades in much of the eastern United States as well as nationwide. This increase is attributed in part to the creation of favorable habitats resulting from conversion of forests to either farmland or habitats characterized by early-successional stages

Figure 7.7. A red fox. (Photograph by Larry Master.)

(Barber 1984). Other contributing factors are reduced hunting pressure and greater use of farm crops as food (Hesselton and Hesselton 1982).

Two wild canids, the red fox (Figure 7.7) and the coyote, have been favored by increased edge and forest fragmentation. Today, the red fox is the most widely distributed canid in the world, occurring in the four continents of North America, Europe, Asia, and Australia (Samuel and Nelson 1982). This fox was probably not native to the eastern deciduous forest but arrived from northerly latitudes of Canada. Red foxes also may have descended from foxes brought to the Southeast by British hunters in the 1750s. Hence, the current success of the red fox in the eastern deciduous forest, as well as elsewhere throughout most of the United States, is attributed partially to the conversion of forest to agriculture.

Like the range of the red fox and in marked contrast to the ranges of most large carnivores, the range of the coyote has expanded since the arrival of Europeans to North America. Currently, the coyote can be found from northern Alaska to Costa Rica (Bekoff 1982). Part of this range extension may be due to the decline in numbers of gray wolves by the early 1900s in the United States. In the 1930s, an "eastern" coyote began to appear in the eastern deciduous forest and now occurs in all eastern states. Replacement of the eastern deciduous forest by agriculture was largely responsible for the recent spread of coyotes into the East.

The Virginia opossum is the only marsupial in North America (Gardner 1982). The opossum is highly adaptable to conditions created by land uses in the eastern United States, such as farming and suburbia. Since the early twentieth century, its distribution has spread from about Maryland and West Virginia northward to about southern Ontario. Future, more northerly range extensions of opossums will probably be limited by winter weather conditions in the northern latitudes.

Fragmentation of forest into a mosaic of small (1 hectare) clear-cut plots of different ages for ruffed grouse habitat provides a variety of habitats and edge for small mammals. Two species in particular, the white-footed mouse and the southern red-backed vole, respond favorably to a range of younger and uncut forest plots within a small area (e.g., 4 hectares) (Yahner 1992).

### *Some Representative Birds Benefited by Edge and Fragmentation*

Before the arrival of Europeans, the status of mallards in the eastern deciduous forest was obscure, but market hunting soon reduced populations along the Atlantic flyway after European settlement (Heusmann 1991). Today, populations have increased throughout the eastern United States. Conversion of forest to farmland coupled with regulated hunting seasons, release of farm-reared birds, and winter feeding by the public are major causes of this increase.

In contrast to mallards, black ducks have been affected negatively by forest fragmentation (Brodsky and Weatherhead 1984). Before the 1900s, these two closely related species were isolated geographically in the eastern United States. Deforestation of the eastern deciduous forest since that time has reduced the availability of breeding habitat for the black duck while providing suitable habitat for the mallard. Hence, mallards have invaded breeding areas of black ducks, particularly on wintering grounds, where courtship and pair bonding generally occur. As a consequence, much hybridization exists between the less common black duck and the more numerous mallard, leading some to believe that the black duck may disappear as a "pure" species.

Populations of certain raptors, such as the great horned owl and the red-tailed hawk, may be on the rise in fragmented landscapes (Figure 7.8). This increase is perhaps a direct result of an increase in major species of prey (Goodrich and Senner 1988). In southeastern Pennsylvania, habitat extensively fragmented by housing developments, parks, small woodlots, and farmland provides suitable habitat for Norway rats,

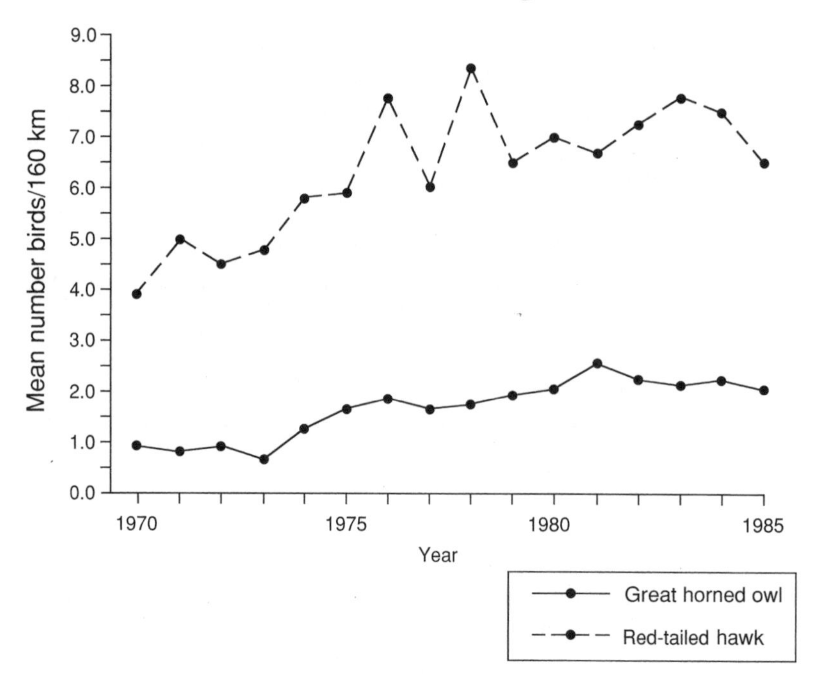


Figure 7.8. Mean number of great horned owls and red-tailed hawks in Pennsylvania, based on Christmas Bird Counts, 1970–85. (Modified from Goodrich and Senner 1988.)

which are an important prey of the great horned owl (Wink, Senner, and Goodrich 1987). An intermixture of forest and farmland is particularly suitable habitat for great horned owls (Morrell and Yahner 1994).

Many songbirds are benefited by increased edge resulting from fragmentation in the eastern deciduous forest. Gray catbirds and common yellowthroats are quite numerous in brushy and edge habitats created by clear-cut stands, transmission-line corridors, or forest irrigation (Chasko and Gates 1982; Yahner 1987, 1993a, 1995b, 1997a; Bramble, Yahner, and Byrnes 1992; Piergallini 1998). The gray catbird has increased over the past few decades in some highly fragmented areas of the eastern United States (Anderson et al. 1981). Conversion of the eastern deciduous forest to farmland or suburbia has had a positive influence on population trends of American crows and blue jays (Wilcove 1988). Populations of blue jays are increasing in suburbia largely because of abundant food provided at bird feeders (Bock and Lepthien 1976; Hickey and Brittingham 1991).

The brown-headed cowbird, which is the only obligate nest parasite of other songbird species in the United States, has increased in numbers in the eastern United States since the early 1900s (Figure 7.9) (Brittingham and Temple 1983; Terbough 1989). Before the 1900s, the cowbird occurred primarily in the plains and the prairies west of the Mississippi River. It was absent from the large unbroken tracts of forest in the East because its feeding habits and social behavior restricted it to open areas. But the distribution of cowbirds moved eastward as the eastern deciduous forest was logged and farmed. Today, cowbird abundance in some forested tracts affected by clear-cutting or transmission rights-of-way in the northeastern United States is relatively low and similar to that in uncut forested stands (Thompson et al. 1992; Welsh and Healy 1993; Yahner 1993a, 1995c, 1997a). Furthermore, perhaps because of farmland abandonment and conversion to forested habitat in the Northeast

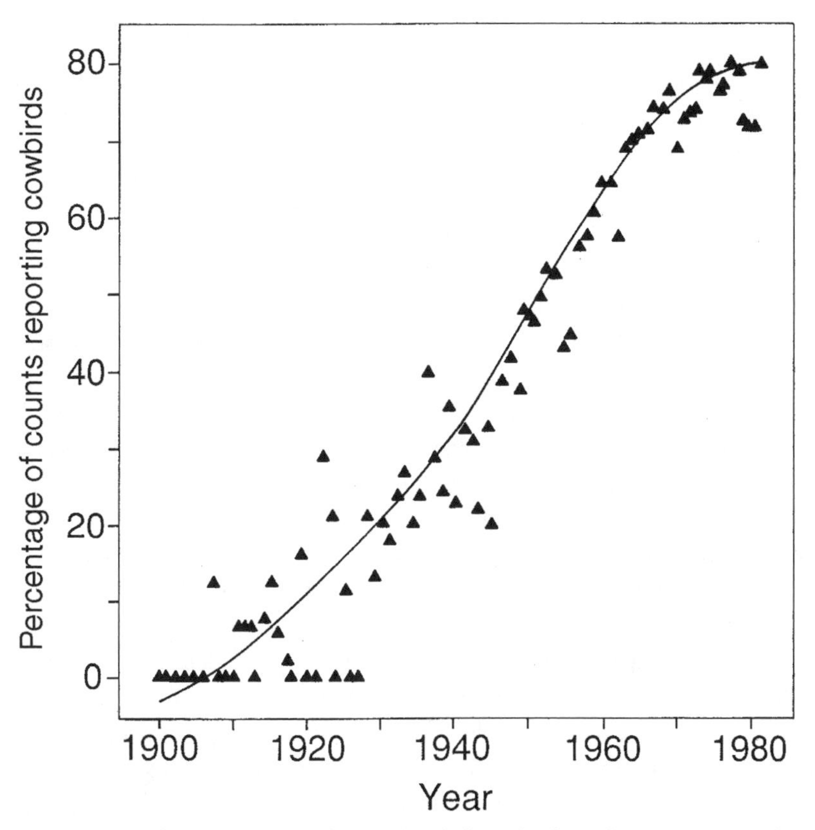


Figure 7.9. Relative increase in brown-headed cowbirds in the eastern United States from 1900 to 1980. (Modified from Brittingham and Temple, 1983.)

over the past few decades (e.g., Litvaitis 1993), cowbird populations have shown regional declines of 3 percent over the period from 1966 to 1994 (Rosenberg and Wells 1995).

In the northeastern United States, cowbirds are associated mainly with agricultural landscapes rather than forested landscapes. For example, cowbird distribution and abundance in Vermont were correlated positively with scattered livestock farms, which were used as foraging sites by cowbirds (Coker and Capen 1995). In central Pennsylvania, cowbird abundance in contiguous forested sites was significantly higher when sites were near (< 1 kilometer) agricultural habitats compared to those near clear-cut stands (Rodewald 2000). Thus, cowbird parasitism on songbirds in forested landscapes may be mitigated by limiting or restricting silvicultural practices near pastures or other agricultural areas (Wood, Duguay, and Nichols 1998).

## Wildlife Species Negatively Affected by Edge and Fragmentation

### *Edges—A Wildlife Management Dilemma*

Despite possible benefits of edges to certain species of wildlife, natural resource managers no longer can assume that the creation of more edge in a forested landscape will always be beneficial to wildlife (Yahner 1988a). In northern New Hampshire, some forest birds, such as red-eyed vireos and hermit thrushes, established fewer territories along edges created by clear-cutting compared to areas farther into the forest; conversely, territories of black-throated blue warblers were more common along edges (King, Griffin, and DeGraaf 1996). Lower abundances of forest bird species along edges may be due to avoidance of high rates of predation along edges, differences in vegetation between forest edges and interiors, or simply the absence of suitable habitat in adjacent clear-cut stands (Gates and Gysel 1978; King, Griffin, and DeGraaf 1996). White-footed mice also were less common along forest edges adjacent to clear-cut stands than within interiors of uncut stands (Yahner 1986b). In central Pennsylvania, gray squirrels and redback salamanders were less abundant along forest edges proximal to farmland compared to forest interiors despite no discernible differences in vegetation between edges and interiors (Derge 1997; Derge and Yahner 2000; Geer 1997). As discussed in chapter 6, edges have reduced moisture levels in surface soil and leaf litter, which may account for low abundances of salamanders along edges (Geer 1997). Presumably, both the abundance and distribution of several species of northeastern amphibians are affected negatively by even-aged management for distances extending at least 25 to 35 meters into the nearby forest (deMaynadier and Hunter 1998).

Abrupt discontinuities in a forested landscape, such as those created by a forest-agricultural field interface, may act as an "ecological trap" for nesting birds (Gates and Gysel 1978). These abrupt, human-made forest edges are recent in origin and are not representative of habitat to which many forest species have adapted. As we have seen in the previous chapter, breeding populations of many forest songbirds have been declining or eliminated in small tracts of the eastern deciduous forest since the late 1970s, particularly those species that winter in the Neotropical forests of Central and South America. Nest predation and parasitism have been suggested as two possible factors affecting populations of these Neotropical migrants (Böhning-Gaese, Taper, and Brown 1993). Furthermore, Neotropical migrants have several biological constraints, such as production of only one brood per year, a late arrival in spring that leaves little time for renesting, and dependency on invertebrate food supplies that may decrease with reduced size of forested tracts.

### *Nest Predation on Forest Songbirds*

In order to better understand the impact of edges on the nesting success of forest birds, at least six potential factors have to be considered, including distance of nests from the edge, characteristics of the edge, type of landscape, size of forested tracts, extent of fragmentation in the landscape, and composition of the predator community. Whether incidences of predation on bird nests are greater along edges than within interiors of forested tracts is not clearly documented (Paton 1994). Some studies using artificial nests in landscapes dominated by farmlands or suburbia have concluded that predation rates in forested tracts may decrease with greater distances from an edge (Wilcove 1985; Andrén and Angelstam 1988). In contrast, no relationship has been found between predation rates on nests and distances to edge in other studies (Ratti and Reese 1988; Esler and Grand 1993), especially those in forest-dominated landscapes (Yahner and Wright 1985; Rudnicky and Hunter 1993b; Yahner, Mahan, and DeLong 1993a).

Logging roads in forested landscapes have been suspected of serving as travel lanes for predators, thereby increasing predation rates on nests of forest birds located near roads (e.g., Askins 1994). However, in a managed forested landscape in central Pennsylvania, predation rates on artificial ground nests were actually greater away (> 50 meters) from roads (58 percent disturbed) than adjacent (1 meter) to roads (45–49 percent disturbed; Yahner and Mahan 1997b); in this same study, predation rates at interfaces of clear-cut and uncut stands were much higher (84 percent disturbed). In highly fragmented landscapes, mammalian nest predators, e.g., raccoons and opossums, may forage as often along forest edges as within forest interiors (Heske 1995). Abundances of several generalist predators, e.g., raccoons, red foxes, and coyotes, are probably associated more with the presence of a high diversity of habitat types (agricultural, grassland, developed sites) rather than location or amount of edge, such as that created by roads (Oehler and Litvaitis 1996).

Nesting success by birds may depend on the characteristics of a particular edge. Predation rates on nests of indigo buntings located near abrupt, agricultural edges, for instance, were nearly twice as high as those placed near gradual edges where plant succession had occurred (Suarez, Pfenning, and Robinson 1997). By comparison, predation rates on both artificial ground and aboveground nests in a managed forested landscape were similar regardless of the degree of plant succession along interfaces of clear-cut and uncut stands (Yahner, Morrell, and Rachael 1989).

Some evidence obtained from artificial nest studies suggests that predation rates are lower in forested landscapes than in agricultural landscapes, irrespective of distances of nests from edges (Bayne and Hobson 1997). Moreover, in forested landscapes affected by clear-cutting, predation rates on nests in relatively wide (≥ 100–150 meters) corridors of uncut forest may be comparable to those on nests in contiguous uncut stands; conversely, predation rates on nests were appreciably higher in isolated uncut stands surrounded by clear-cut stands (Yahner and Ross 1995; Vander Haegen and DeGraaf 1996; Yahner and Mahan 1996b).

A general trend found in predation rates on avian nests in agricultural and suburban landscapes is that rates are higher in smaller forested tracts than in larger tracts (Wilcove 1985; Donovan et al. 1995). In an experimental study of artificial nests using fresh Japanese quail eggs, for example, predation rates were higher in smaller woodlots than in larger woodlots that were surrounded by farmland and suburbia (Wilcove 1985) (Figure 7.10). In woodlots of less than 100 hectares, predation rates were at least 25 percent but were considerably lower in larger tracts (< 10 percent). In the largest forested tract (represented by the Great Smoky Mountains National Park), the predation rate was only 2 percent. This low rate of predation on nests in the Smokies may be comparable

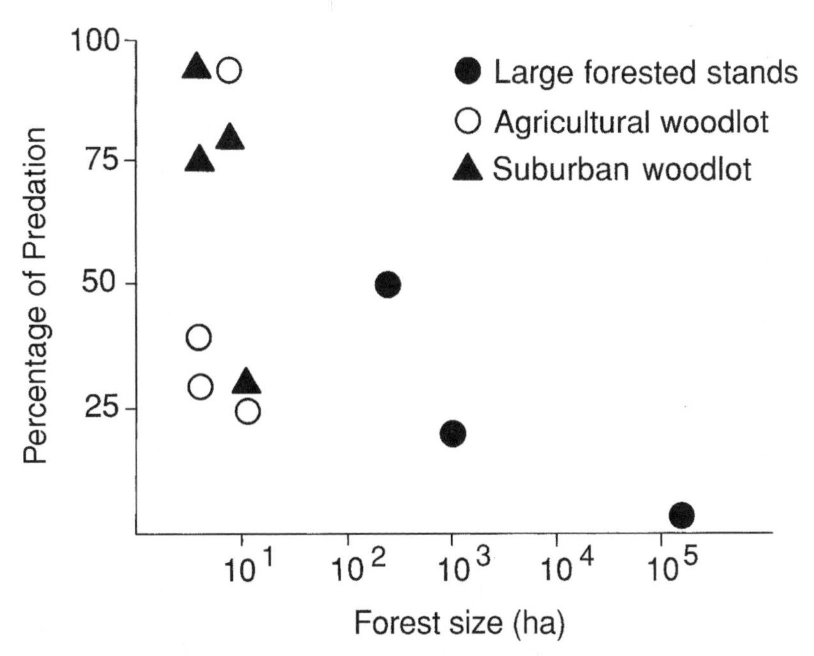


Figure 7.10. Predation rates on avian nests in forested tracts of different sizes in landscapes dominated by farmland and suburbia. (Modified from Wilcove 1985.)

to those historically found on bird nests before extensive fragmentation of the eastern forest by Europeans. Higher predation on nests in suburban forested tracts compared to rural tracts may reflect higher densities of predators associated with suburbia, such as blue jays and common grackles. In southeastern Pennsylvania, predation on wood thrush nests also differed with size of woodlots surrounded by agricultural lands; 56 percent of the nests in smaller woodlots (< 80 hectares) were destroyed by predators compared to only 22 percent of the nests in larger woodlots (> 100 hectares) or contiguous forest (> 10,000 hectares) (Hoover, Brittingham, and Goodrich 1995). In contrast, studies of predation on artificial avian nests in a forest-dominated landscape have indicated that nest loss does not vary either with the size of large, uncut tracts (Leimgruber, McShea, and Rappole 1994) or among various sizes of uncut and clear-cut tracts (Rudnicky and Hunter 1993b).

Nest predation may increase in a forest-dominated landscape with greater amounts of forest fragmentation. For instance, in an experimental study of artificial nests using fresh, brown chicken eggs, only 9 percent of the nests were disturbed by predators in a mature, uncut forested tract in a ruffed grouse habitat management area in central Pennsylvania (Yahner and Scott 1988). Conversely, 19 percent of the nests were lost to predators in a forested tract with 25 percent clear-cutting, and 50 percent of the nests were disturbed in a tract with 50 percent clear-cutting; nest losses were attributed mainly to blue jays and American crows, whose numbers generally increased with greater amounts of forest edge. But on this same study site several years later, a moderate increase in the extent of forest clear-cutting resulted in less favorable habitat for jays and crows (Yahner, Mahan, and DeLong 1993). Overall predation rates, therefore, dropped at least 31 percent with declines in abundance of these avian predators on the management area.

Predation rates on bird nests in a given landscape, whether forested or agricultural, cannot be fully understood without considering the abundance, distribution, and dynamics of the predator communities (Yahner 1996). For instance, avian predators may have the major effect on bird nests in forested landscapes, whereas mammalian predators are the principal cause of nest predation in agricultural landscapes (Picman 1988; Yahner and Scott 1988; Haskell 1995). The type of predator (bird versus mammal) apparently also tends to differ with the height of nests. Aboveground nests usually are preyed upon more often by birds than mammals, but the reverse may be true of ground-level nests (Yahner, Morrell, and Rachael 1989; Bayne and Hobson 1997).

### *Nest Parasitism by Cowbirds on Forest Songbirds*

Historically, the cowbird probably parasitized open-cup nests of about fifty species of birds compared to perhaps two hundred species today (Mayfield 1965). Because the cowbird is an edge species and many forest songbirds nest near edges (Gates and Gysel 1978), nests of forest songbirds are more likely to be found and parasitized by cowbirds than nests farther from edges. Some of these species have not evolved defenses against cowbird parasitism and are termed "acceptors," because they do not recognize cowbird eggs and will raise cowbird young that hatch. Chestnut-sided warblers (Figure 7.11) and wood thrushes are examples of acceptor species; the gray catbird, on the other hand, is a "rejector" species and will evict cowbird eggs deposited in its nest.

In general, the effects of cowbird parasitism on the decline of forest birds may not be as dramatic in the East as in the Midwest. An examination of nest records of wood thrushes from the Cornell Laboratory of

Figure 7.11. A chestnut-sided warbler nest with five eggs. The two larger and well-speckled eggs are those of a cowbird. (Photograph by author.)

Ornithology, Cornell University, over a thirty-year period (1960–90) showed that numbers of nests parasitized by cowbirds increased from east to west; only 15 percent of wood thrush nests were parasitized in the northeastern states compared to more than 42 percent of nests in the midwestern states (Hoover and Brittingham 1993). In New York, for instance, only 8 percent of wood thrush nests were parasitized by cowbirds (Hahn and Hatfield 1995). In southeastern and central Pennsylvania, 9 percent and 10 percent of wood thrush nests were parasitized by cowbirds, respectively (Hoover, Brittingham, and Goodrich 1995; Yahner and Ross 1995). In central Pennsylvania, only 3 percent of nests of several species of songbirds combined were parasitized in forests managed by clear-cutting (Yahner 1991).

Susceptibility of forest songbirds to cowbird parasitism varies not only among bird species and geographic area but may differ with distances of nests from an edge. In a study of parasitism by cowbirds on a number of forest songbird species in Wisconsin, 48 percent of 105 nests were parasitized, and rates of parasitism were significantly higher on nests closer to edges (Brittingham and Temple 1983). For example, 65 percent of nests within 100 meters of edges were parasitized compared to only 3 percent beyond 300 meters. However, in another study of cowbird parasitism in Wisconsin, no relationship was found between distances of acadian flycatcher nests from edges and rates of parasitism (Bielefeldt and Rosenfield 1997). Furthermore, in Pennsylvania, rates of parasitism by cowbirds on ovenbird nests did not differ with distances from woodlot edges (Giocomo 1998). A study of nest parasitism in Vermont found a much different relationship between edge and parasitism, with only 7 percent of the nests of several bird species parasitized by cowbirds along forest edges compared to 32 percent of the nests parasitized in forest interiors (Hahn and Hatfield 1995).

As with the relationship between rates of parasitism by cowbirds and distances of nests from edges, conflicting results have been obtained with rates and woodlot size. For example, rates of nest parasitism on nests of ovenbirds, red-eyed vireos, and wood thrushes were higher in forested tracts (averaging 550–675 hectares) than in contiguous forests (> 26,794 hectares) in three midwestern states (Minnesota, Missouri, and Wisconsin) (Donovan et al. 1995). On the other hand, no difference was found with rates of parasitism on wood thrush nests in forested tracts (9–127 hectares) versus contiguous forest (> 10,000 hectares) in southeastern Pennsylvania (Hoover, Brittingham, and Goodrich 1995). Thus, these studies of nest parasitism since the mid-1990s have given us a different perspective of the effects of cowbirds on the nesting success of forest birds in relation to either distance from edges or size of forested tracts—one which differed from the perspective of the late 1980s and early 1990s (Yahner 1995a). Future studies will be necessary to give natural resource managers a better understanding of the impact of cowbirds on the nesting success of birds in the eastern deciduous forest.

In Illinois, which is much less forested than many states in the East (Figure 1.2), forested tracts occur as islands in an agricultural landscape; these tracts are generally small and are therefore essentially all edge habitat with little or no forest-interior habitat. Hence, forest songbirds, such as wood thrushes, are very exposed to cowbird activity. At least 70 percent of the wood thrush nests in certain areas of Illinois were parasitized, suggesting that the future of wood thrush populations in Illinois is in jeopardy (Robinson 1992). A subsequent study that was conducted a few years later on the same study areas in Illinois, however, reported only a 25 percent parasitism rate on wood thrush nests (Bollinger and Linder 1994).

A couple of recent studies have provided evidence that actual or potential rates of cowbird parasitism on nests of forest songbirds may differ within a region and often with characteristics of the landscape on a broad scale (Yahner 1995a). For instance, in Delaware, rates of nest parasitism were very different between two sites separated by only 1.2 kilometers (Roth and Johnson 1993). In Vermont, cowbirds seldom used small forest openings (approximately 4 hectares ), whereas cowbirds typically were attracted to larger openings (> 9 hectares) associated with many livestock areas within the vicinity (< 7-kilometer radius) (Coker and Capen 1995). In the Midwest, rates of cowbird parasitism of several bird species, e.g., indigo buntings, ovenbirds, and red-eyed vireos, were greater in those wooded study sites containing relatively low amounts of forest cover in the general area (10-kilometer radius) (Robinson et al. 1995).

Cowbird parasitism may have serious consequences for the future status of endangered or threatened songbirds. For instance, the Kirtland's warbler, an acceptor species, has federal endangered status. Some wildlife biologists have contended that without an ongoing program of cowbird removal in nesting habitat of the Kirtland's warbler, the remaining two hundred individuals in Michigan will go extinct (Mayfield 1977) (see Figure 7.4).

In conclusion, edges resulting from forest fragmentation can be harmful to the long-term distribution and abundance of a variety of Neotropical migratory songbirds and other wildlife in the eastern deciduous forest. Evidence suggests that, at least in the Midwest, nests of forest songbirds located in forested tracts even as large as 500 hectares may not provide secure habitats from high rates of predation and parasitism. Hence, natural resource managers and land-use planners should commit to regional conservation efforts that maintain, restore, and preserve large forested tracts for the future benefit of forest wildlife.

## Program Developments for the Conservation of Migratory Birds

The future conservation of Neotropical migratory birds will require a concerted and joint effort involving many countries. A step in this direction has been taken by the National Fish and Wildlife (NFW) Foundation, a not-for-profit organization chartered by Congress. The NFW Foundation proposed a major initiative in 1990 to begin a comprehensive program concentrating on research, monitoring, and management of Neotropical migrants in North America (Finch 1991). This massive conservation program, called Partners in Flight, involves federal agencies, academics, and private-sector scientists (Hagan 1992). In 1993, about fourteen federal agencies, all fifty states, and at least twenty nongovernmental groups participated in this program (Martin 1993). A goal of the Partners in Flight program is to identify assemblages of high-priority Neotropical species and focus on the conservation of their habitats (Pashley, Hunter, and Carter 1992). The NFW Foundation recommended that research and conservation efforts be centered especially in countries containing the majority of wintering habitat for these birds, including the Bahama Islands, Belize, Cuba, the Dominican Republic, Guatemala, Haiti, Honduras, and Mexico. Cooperative research, monitoring, and manipulation of habitat would be implemented simultaneously in breeding and wintering areas to help minimize the loss of biodiversity in both temperate and tropical forests.

The Partners in Flight program also works at a more regional level. For example, the Northeast Region of the Partners in Flight program has examined the status of Neotropical migratory landbirds in thirteen northeastern states, extending from Virginia to Maine (Rosenberg and Wells 1995). The Northeast Region also has developed three action items: (1) identify and rank physiographic provinces and key habitats in terms of their importance to birds, (2) identify centers of population abundances for priority bird species, and (3) develop strategies for the conservation of important habitats or areas for priority species. Information obtained by this regional program of Partners in Flight will have tremendous value to state and regional planners concerned about the long-term conservation of Neotropical migratory birds in the eastern deciduous forest.

The conservation of breeding grounds for Neotropical migrants in North America will not solve completely the international problem of saving these songbirds. However, these efforts send a signal to tropical nations that conservation agencies in North America are committed to solving this problem (Maurer and Heywood 1993). Both national and international efforts in the conservation of our migratory songbirds are critical steps in the right direction. Databases and long-term monitoring projects will be necessary to develop international conservation efforts for Neotropical migrants (Kattan, Alvarez-López, and Giraldo 1994). Hence, the conservation of these birds stands out as an issue unmatched in the history of American wildlife conservation because of the level of commitment by a diverse group of parties and nations (Hagan 1992). This effort is patterned after the North American Waterfowl Management Plan among the United States, Canada, Mexico, and private conservation organizations (M. W. Weller, Texas A & M University, personal communication). The conservation of Neotropical migratory birds is an excellent example of the growing concern for worldwide loss of biodiversity and of conservation efforts aimed at mitigating its loss, which is a principal focus of the next chapter.

# 8. BIODIVERSITY AND CONSERVATION

## What Is Biodiversity?

I have referred to biological diversity, or simply biodiversity, several times earlier, but it may mean different things to different readers and, therefore, be somewhat ambiguous (Yahner 1993b). In fact, eighty-five different definitions of the term biodiversity have been found based on a relatively recent survey (DeLong 1996)! Criticism has been leveled at some definitions of biodiversity because the term may be a catchall for everything biotic and abiotic and, thus, is too cumbersome from a pragmatic, conservation perspective (Lautenschlager 1997). A more focused definition of biodiversity is the variety within and among biotic communities at a given site or area; note that this definition excludes the abiotic components of a site or area (DeLong 1996).

A rather inclusive and generally accepted definition of biodiversity is the variety of life and its processes (Keystone Center 1991; Nigh et al. 1992; The Wildlife Society 1993; Noss and Cooperrider 1994). Using this definition, biodiversity has been described at four interrelated levels: genetic diversity, species diversity, community and ecosystem diversity, and landscape diversity (Figure 8.1). More specifically, biodiversity is the variety and the abundance of species, their genetic diversity, and the communities, ecosystems, and landscapes in which these species occur; in addition, biodiversity refers to ecological structures, functions, and processes at each of these four levels (Society of American Foresters 1992).

Genetic diversity, which represents the "smallest" scale of biodiversity, is the variation in genetic makeup of individuals of the same species

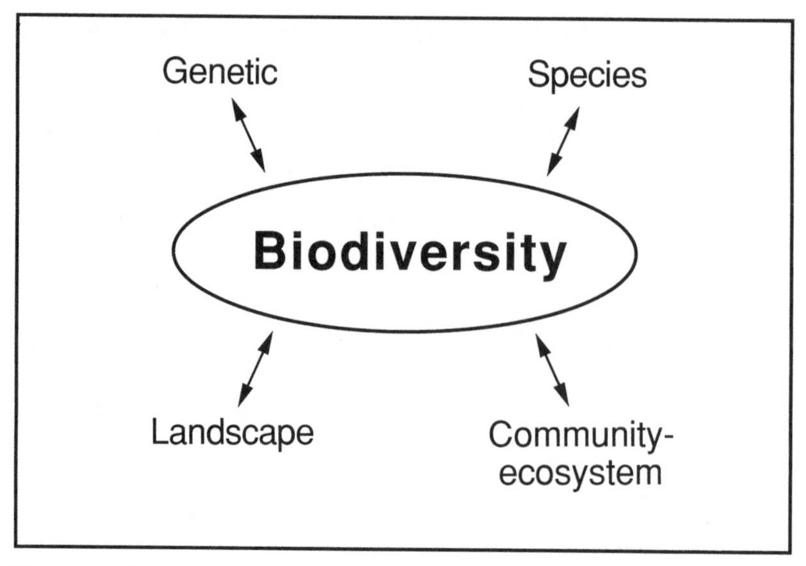


Figure 8.1. Four interrelated levels of biodiversity.

within a population or group of populations in a prescribed geographic area (Nigh et al. 1992). Sufficient genetic variation is important because it may better ensure that a population or a species can adapt to changes in the environment, such as a climatic change, and enhances its chances of long-term survival.

The genetic variation of some common wildlife species, like black bears and yellow-bellied slider turtles, has been studied in recent decades (Walthen, McCracken, and Pelton 1985; Scribner et al. 1986). Information on the genetics of common species certainly will have value in the future to biodiversity conservation in the eastern deciduous forest. In contrast, an understanding of the genetics of endangered species, such as red wolves and Florida panthers, is necessary for their survival. As another example, the success of recovery programs for gray wolves in the Rocky Mountains is contingent on ensuring that high genetic variation is present in transplanted individuals (Forbes and Boyd 1996); moreover, genetic data are valuable in determining the extent to which wolves from Canada have dispersed into populations of wolves transplanted into Glacier National Park and elsewhere in the central Rockies.

For many years, we have preserved the existing genetic diversity of species with economic importance to humans, such as farm crops, farm animals, biomedically important animals, and microorganisms used in food, environmental, and biomedical industries, using genome resource banks (Wildt et al. 1997). Today, the strategy of genome resource banks is being implemented in conservation plans for high-priority species worldwide, such as chimpanzees and giant pandas. The success of the Scottish team in cloning the domestic sheep known as Dolly, however, has raised some ethical issues surrounding the use of cloning to offset the loss of genetic diversity in endangered species conservation (Bawa, Menon, and Gorman 1997).

Species diversity is the variety of living organisms. Several decades ago, Aldo Leopold (1933) recognized the importance of this concept to natural resource management. Today and in coming years, conservation efforts both worldwide and throughout North America will continue to be formulated using this concept. Species diversity can be calculated in various ways, ranging from compiling a simple list of the number of different species, termed species richness, to more complicated measures that are based also on the abundance of individual species. The number of species and abundance of each on a localized basis is called alpha species diversity; this measure has been emphasized by wildlife managers and researchers over the years (Samson and Knopf 1982). Its determination requires considerable time and effort but is feasible, for example, when measuring breeding bird diversity in mixed-oak and aspen stands of various ages since clear-cutting (e.g., Yahner 1986, 1993a).

Gamma species diversity, on the other hand, is the number of species over a broad geographic area, with little or no reference to the abundance of each species. In the past, gamma diversity largely was ignored by natural resource managers, yet it may be the most pragmatic measure of species diversity in certain situations (Samson and Knopf 1982).

Gamma diversity has become especially important in recent years with the passage of legislation that focuses on the biodiversity of large land tracts, such as national forests (National Forest Management Act of 1976). Consider, for instance, a natural resource agency that is mandated to manage for all plant and animal life rather than for just a few species (e.g., U.S. Forest Service). The agency may be able to develop an extensive list of species on public lands but be unable to gather information on the abundance of each species because of limited monies, time, and labor. However, ambitious monitoring of wildlife can provide enormous amounts of valuable information. A good example of such a monitoring effort is the breeding bird survey, which is an ongoing cooperative program sponsored jointly by the U.S. Fish and Wildlife Service and the Canadian Wildlife Service (Robbins, Bystrak, and Geissler 1986). The breeding bird survey attempts to estimate population trends of birds that nest in North America north of Mexico and migrate across national boundaries. This important database, obtained by an impressive number of volunteer birders, has shown, for instance, that populations of house sparrows have declined in the eastern United States, whereas the exotic house finch, introduced on Long Island, New York, in 1942, has increased throughout the East an average of 21 percent per year from 1966 through 1979 (Robbins, Bystrak, and Geissler 1986). Some other important trends shown in bird life in the East by the breeding bird survey are increases in black-capped chickadees, house wrens, and American robins and declines in populations of northern flickers; declines in flickers are attributed partially to competition with European starlings for nesting holes (Ingold 1994).

Statewide monitoring of wildlife by volunteer birders also has been undertaken in recent years to gain information on gamma diversity. A case in point is the Pennsylvania Breeding Bird Atlas Project, which involved numerous enthusiastic and dedicated professionals and amateur birders (Brauning 1992). These birders spent many hours in the field listing breeding bird species, as well as collecting data on relative abundance and population trends of each species.

Community or ecosystem diversity is the variety of communities or ecosystems that occur over a broad geographic region or landscape (Nigh et al. 1992). This measure of diversity is enhanced when a spectrum of all possible habitats, ranging from upland forests to bottomland wetlands, is maintained within a given forested landscape. Moreover, preservation of critical habitats, such as rocky outcrops and caves within an extensive forest, may be keys to long-term conservation efforts for certain species of special concern in the eastern deciduous forest. For instance, since perhaps the 1960s or earlier, populations of Appalachian woodrats in several states, including New Jersey, New York, and Pennsylvania, have experienced dramatic declines in numbers for reasons that have not yet been determined completely by wildlife biologists (e.g., Balcom and Yahner 1994, 1996) (Figure 8.2).

Landscape diversity, or diversity of landscapes within a given geographic region, represents the "largest" scale of biodiversity, excluding the biosphere. This scale incorporates numerous interacting communities and ecosystems, natural and human-induced disturbances, and, optimally, viable populations of wide-ranging birds and mammals (Noss 1992). We can perhaps view large tracts of land, such as state and federal public lands in the East (e.g., the Allegheny National Forest) as fitting into this scale of biodiversity. Management of landscape diversity can be

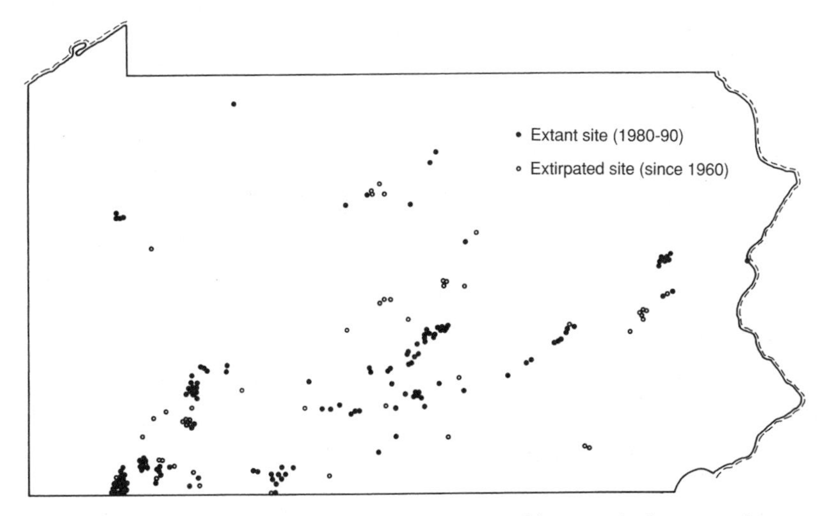


Figure 8.2. Known extant and extirpated sites used by Appalachian woodrats in Pennsylvania. (Modified from Balcom and Yahner 1994.)

very important to the conservation of biodiversity at the regional level and to ecosystem management (Petit, Petit, and Martin 1995). We will discuss ecosystem management later in this chapter.

## Maintenance of Biodiversity Is a Concern

Biodiversity is being recognized increasingly as an essential, nonrenewable natural resource, much in the same way we view tropical forests and stratospheric ozone (Daily and Ehrlich 1992). Moreover, environmental and economic benefits of biodiversity in the United States alone have been valued at $300 billion per year (Pimental et al. 1997). Natural resource managers are concerned about the maintenance of natural biodiversity because of the realization that species are being lost at an unprecedented rate, particularly in the tropics, as a result of habitat disturbance and fragmentation (Miller 1990; Ehrlich and Wilson 1991). An annual rate of extinction of twenty thousand to fifty thousand species of flora and fauna (assuming a conservative ten million total species), or 55 to 150 species per day, is predicted by the beginning of the twenty-first century! Historically, about one species of bird or mammal was lost per hundred years between 10,000 B.P. and 1975; in recent years, however, this rate has been estimated at one species per year. An estimated 10 percent of the world's plant species, or about sixty thousand species, are threatened with extinction in the near future; this statistic raises a serious issue because plants are the cornerstones of ecosystems (Ellstrand and Elam 1993).

Consider for a moment that each species is a unique entity and took perhaps two thousand to ten thousand generations to evolve; its extinction then can be likened to the loss of a priceless piece of art. Dr. Paul P. Ehrlich once said that if a lost species were instead the Mona Lisa (which took three years to paint and not two thousand to ten thousand generations), both the art community and society would feel great anguish. Unfortunately, many species facing imminent extinction have not yet been classified or even discovered. Ironically, this scientific task will likely go uncompleted before a large fraction of species are lost because of a serious worldwide shortage of professional systematists who have the expertise to identify species accurately, particularly those found in tropical forest ecosystems (Wilson 1988). Hence, the training of international personnel to conserve biodiversity in developing countries must be an important conservation goal in the immediate future (Wemmer et al. 1993). The conservation of biodiversity not only requires scientists trained in basic biological and ecological principles but also those versed in taxonomy, systematics, and basic natural history of organisms. Today, there is an alarming decline in the teaching of these latter skills in museums and universities worldwide (Noss 1996).

Another important reason why society should be concerned about the conservation of biodiversity is that it is the foundation of the earth's ecosystems. It is essential to the health and the productivity of forests (Society of American Foresters 1992). Species serve many purposes in an ecosystem: Plants produce oxygen and regulate water supplies, and animals serve as herbivores, predators, pollinators, and scavengers. Therefore, a natural diversity of species indicates a healthy environment (Karr 1990; Probst and Crow 1991). From an ecological standpoint, even the loss of a single species could have damaging effects on an ecosystem and its remaining biota in ways that are not yet understood or predictable, especially if that species is a key "link" (keystone species) between other organisms in the ecosystem.

Biodiversity provides sources of valuable renewable natural resources, such as food for humans (Society of American Foresters 1981). In eastern Canada, Native Americans gathered about 175 foods and 52 beverages from forest plants alone (Arnason, Hebda, and Johns 1981). The bulb of the wood lily was a food source to Native Americans in the eastern deciduous forest (Niering and Olmstead 1983). About 90 percent of the world's domestic plants have been derived from a few wild species that originated principally from tropical forests. Of the 250,000 species of plants in the world, about twenty thousand species (8 percent) have been used by humans as food (Pimental et al. 1997). However, just over one hundred species (0.04 percent) contribute to 90 percent of the world's plant food supply (Prescott-Allen and Prescott-Allen 1990). Clearly, the conservation of plant biodiversity (species and genetic) is essential to feeding the world's growing human population. The recognition that plant biodiversity must be preserved was a major outcome of the 1992 Earth Summit in Rio de Janeiro (Barton 1992).

Plants and animals not only provide major sources of food but also contain important chemicals for prevention and possible cure of human illnesses. Many plant and animal products of the world's forests have been used for medicinal purposes by indigenous peoples for centuries (Bennett 1992). Native Americans from eastern Canada derived four hundred medicines from forest plants (Arnason, Hebda, and Johns 1981). In the eastern deciduous forest, Native Americans used the wildflower yarrow for medicinal purposes (Niering and Olmstead 1983). In the United States, about 25 percent of today's prescription drugs contain ingredients from wild species of plants (Miller 1990). Recently, the Pacific yew, which occurs in the old-growth forest of the Pacific Northwest, has been found to contain taxol, a chemical that can be used to treat various types of human cancer (Nicholson 1992).

The economic return of plant-based drugs is estimated at about $36 billion per year in the United States and more than $200 billion per year worldwide (Pimental et al. 1997). Hence, future researchers will continue to seek plant and animal derivatives that hold the key to treating or eliminating dreaded diseases worldwide; perhaps some of these derivatives will be found someday in the eastern deciduous forest. Furthermore, the pharmaceutical industry's exploration during the next four decades for animal and plant derivatives useful to human medicine will have a major, positive effect on preserving the world's biodiversity (Eisner and Beiring 1994). The hope is that the present value of medicines combined with those of other sustainable nontimber products derived from forests worldwide will help ensure sound conservation and management of forests well into the future (Balick and Mendelsohn 1992).

Biodiversity has aesthetic and recreational benefits for outdoor and wildlife enthusiasts. Birding, ecotourism, outdoor photography, and other forms of nonconsumptive recreation, as well as the more traditional wildlife activities such as hunting and fishing, are enjoyed by more than 50 percent of the people in the United States annually. Outdoor recreation contributed nearly $60 billion to the U.S. economy in 1991. Birding, in particular, is one of the most rapidly growing wildlife recreational activities in North America, with at least twenty to thirty million participants annually (Kellert 1985); birders in North America spend at least $20 billion per year enjoying this pastime (U.S. Fish and Wildlife Service 1982).

In 1996 alone, 77 million U.S. residents over 16 years old participated in some type of wildlife-related recreational activity (U.S. Department of the Interior 1996b). These included more than 35 million people who fished and fourteen million who hunted; total expenditures by these two types of wildlife recreationists were nearly $72 billion in 1996. In that same year, an additional sixty-three million people observed, fed, or photographed wildlife, which added more than $29 million to the U.S. economy. Various types of outdoor recreation can have a pronounced effect on local economies: For example, hunters at the Delaware Water Gap National Recreation Area, which is along the Pennsylvania and New Jersey borders, spent nearly $1.4 million during the 1987–88 hunting seasons (Strauss et al. 1989).

Conservation of biodiversity is also of concern for legal reasons, with the passage of legislation such as the National Environmental Policy Act of 1969, the Endangered Species Act (ESA) of 1973, and the National Forest Management Act of 1976 (Salwasser 1990). The ESA, in particular, has extended legal rights to nonhuman species (Karr 1990). Perhaps equally important, it has heightened an awareness of the importance of preservation of nontraditional (nonconsumptive) species of wildlife, such as amphibians, reptiles, plants, and their critical habitats (Galligan and Dunson 1979; Moyle and Williams 1990; Wyman 1990) (Figure 8.3). From 1985 to 1991, 68 percent of the species listed as endangered or proposed for listing in the United States were plants, compared to 19 percent vertebrate and 18 percent invertebrate species (Wilcove, McMillan, and Winston 1993).

## Biodiversity and Endangered Species Legislation

Since its passage in 1973, the ESA has been the major foundation for wildlife conservation in the United States by mandating the protection and the conservation of endangered and threatened species and the habitat on which these species depend (Heinen 1995; Carroll et al. 1996). Approximately 80 percent of the taxa listed from 1985 through

Figure 8.3. The timber rattlesnake is a nonconsumptive species in need of protection. (Photograph courtesy of the School of Forest Resources, Pennsylvania State University.)

1991 were full species, 18 percent subspecies, and 2 percent populations (Wilcove, McMillan, and Winston 1993). The average population size for animals at the time of listing was about one thousand individuals compared to only 120 for plants.

Each species listed under the ESA must be the subject of a recovery plan consisting of three parts: natural history of the species, actions needed to reduce threats to the species, and a schedule of recovery (Heinen 1995). Of 344 recovery plans examined in the early 1990s, 306 (89 percent) were intended for a single species, with only 26 percent for more than one species and 12 percent (< 4 percent) for a keystone species (Tear et al. 1995). Thus, recovery and subsequent protection of endangered and threatened species can be an expensive and slow process if the process continues to proceed on a species-by-species approach (Heissenbuttel and Murray 1992). Furthermore, estimates are that it will take nearly fifty years and more than $4.6 billion to recover these wildlife.

Because time and monies required to recover and protect endangered and threatened species are considerable, the ESA's goals were amended in 1988 to conserve the ecosystems upon which these species depend

(Carroll et al. 1996). This goal will be achieved by using a proactive "ecosystem approach" rather than a "single-species approach." The single-species approach has tended to be reactive by responding only when the range of a species and its population size have been seriously reduced (Franklin 1993, 1994; Carroll et al. 1996). The ecosystem approach includes the identification of habitats and communities that are being reduced or degraded, followed by policies and actions that prevent further loss and degradation. Moreover, the ecosystem approach attempts to deal with the conservation of all species, thereby better ensuring that the ecosystem continues to function properly (Walker 1995). Ecosystem-level efforts for biodiversity conservation can reasonably be achieved under the existing ESA (Miller 1996).

Implementation of legislation for the protection of biodiversity or endangered species, however, has created (and probably will continue to create) some bureaucratic problems. Protection of any endangered species generally has been viewed by some parties as an attempt to obstruct progress (Karr 1990). For example, the goals of the ESA were challenged by development of the Tellico Dam in Tennessee, which endangered the snail darter, construction of an astronomical complex that threatened the Mt. Graham red squirrel, and logging in the Pacific Northwest old-growth forests, which put the northern spotted owl at risk (Karr 1990). Interest groups opposed to the ESA have asserted erroneously that the total administrative costs of the ESA have been in the billions of dollars while benefiting only a few species; the total costs of the ESA have actually been only around $700 million while improving the status of 40 percent of the species listed as endangered (Heinen 1995).

Despite the ESA's bureaucratic, economic, or other problems, most natural resource managers are committed to the conservation of biodiversity. Biodiversity and endangered species conservation in the future will require that recovery plans involve public education, public consultation, and perhaps compensation to people adversely affected by conservation efforts (Heinen 1995). In the past, public education usually has been recommended in recovery plans for endangered and threatened species, but consultation with the public has virtually been ignored (Tear et al. 1995). Because less than 10 percent of the species protected by the ESA are on federal lands, effective strategies must be developed in the future to conserve species on private lands and to enlist the support of private property owners (Bean and Wilcove 1997). Also, today and into the future, the conservation of biodiversity and endangered species will depend to some extent on state efforts; all states and Puerto Rico have wildlife programs that benefit nongame and endangered species (Swimmer, Manor, and Gooch 1992).

A comprehensive, national strategy for biodiversity conservation will require sound scientific information on biodiversity from genetic to landscape levels, the availability of both protected and multiple-use areas, and a commitment to ecological and species restoration (Blockstein 1995). The importance of biodiversity conservation, however, extends beyond the borders of North America. It also has tremendous international significance, particularly since the signing of the Convention on Biological Diversity at the Earth Summit in Rio de Janeiro in 1992 (see http://www.biodiv.org). Thus, we should all concur that a strengthening and reauthorization of the Endangered Species Act and initiatives developed by the Convention of Biological Diversity are vital to the future conservation of biodiversity in the eastern deciduous forest and throughout the world (Murphy et al. 1994; Roy 1995). We owe to our children and their children a world that is rich in plants and animals, and the environmental community must continue to strive for the effective preservation of biological diversity on all scales, from genetic to landscape (Blockstein 1990; Hunt and Irwin 1992).

## How Much Diversity Is Enough?

How much diversity is enough? Should all diversity be preserved worldwide, or even in the eastern deciduous forest, including native and exotic species, regardless of costs in terms of monies, time, and labor? Because extinction is a natural phenomenon, why should we be concerned about the California condor, whose numbers have dwindled to just a few despite an enormous conservation effort (Matthews 1991)? Furthermore, should we be concerned about the extirpation of Appalachian woodrats in New Jersey and New York (Balcom and Yahner 1994, 1996)? A life-centered (biocentric) view of diversity considers all species equally valuable, and the loss of one species is as tragic as the loss of any other species (Miller 1992). Given this view, biodiversity should be maximized, and all efforts to save an endangered, threatened, or rare species are particularly attractive because extinction is irreversible. However, a focus on saving ecosystems or habitats containing an assemblage of species, termed an ecosystem-centered (ecocentric) approach (Miller 1992), will have a greater payback for natural resource managers.

The question How much diversity is enough? will continue to linger and is difficult to address from a management perspective. If efforts are directed primarily at managing habitat to achieve maximum biodiversity, we may on occasion run the risk of detrimentally affecting endangered, threatened, or rare species. For example, although grassland habitats in the northeastern United States have relatively few bird species compared to forests, many of these grassland species are experiencing serious population declines across their range. If conservation of grasslands is neglected, future populations of northern harriers, grasshopper sparrows, and other bird species of special concern in grasslands interspersed within the eastern deciduous forest could be in serious jeopardy (Vickery 1991; Yahner and Rohrbaugh 1996, 1998).

Small-scale forest-management practices, such as clear-cutting, can increase species diversity on a localized basis in the eastern deciduous forest (Welsh and Healy 1993; Yahner 1986a, 1993a, 1997a). Yet whenever edge habitat is created, edge or exotic species may benefit at the expense of forest-interior or rare species (Samson and Knopf 1982). Forest clear-cutting, for instance, would not be recommended near nest sites of northern goshawks; this hawk is a forest-interior species and is considered "at risk" in Pennsylvania (Kimmel and Yahner 1994) (Figure 8.4). Thus, a rule of thumb that higher species diversity is always desirable or better should be replaced with one that calls for natural resource managers to examine areas and their corresponding assemblages of species closely and on a case-by-case basis to best achieve conservation goals.

In summary, conservation of biodiversity is not simply maintenance of the maximum number of species at whatever cost. At times, our focus may be on perpetuating a keystone species or a given assemblage of species at a local scale, such as within a particular plant community. Other situations will require that we ensure genetic diversity, such as when dealing with a species of concern. But these species-specific or site-specific efforts, when taken alone, will lead to shortsightedness and eventually be counterproductive unless we also "think big" on both time and spatial scales (Hunt 1991). Conservation of biodiversity will require strategic planning that considers all species and habitats in the long term and over a spatial scale that ensures the presence of corridors and other landscape linkages to mitigate the effects of habitat fragmentation and loss (Kellert 1995). Efforts to achieve biodiversity conservation on these different scales in the future will be challenging and dictated by biological concerns in conjunction with ethical, economic, and other factors (Noss 1992; Kellert 1995; Carey and Curtis 1996).

Figure 8.4. Active nest site of a northern goshawk, which is classified in Pennsylvania as a species "at risk." (Photograph by J. Timothy Kimmel, School of Forest Resources, Pennsylvania State University.)

## Biodiversity, Sustainability, and Ecosystem Management

Maintenance or restoration of biodiversity is essential to sustain both the commodity and noncommodity values of the eastern deciduous forest (Society of American Foresters 1981; Probst and Crow 1991). Traditionally, foresters have regarded sustainability of forests solely as a level of timber harvest or management that could be maintained indefinitely. This early view of sustainability of forests has been replaced by a more contemporary one that can be defined as the long-term capacity of ecosystems to produce values for society (Levin 1993). This view is concerned with human impacts on the biodiversity of flora and fauna and with relationships between biodiversity and ecosystem processes. The contemporary view of sustainability deals better not only with the more prominent species in the forest, such as birds, mammals, and trees, but also with biodiversity of less conspicuous species, such as microorganisms in the soil, invertebrates in the leaf litter, and herbaceous plants on the forest floor (Society of American Foresters 1981; Franklin 1997).

The concept of ecosystem management as it has emerged today is in response to concerns for biodiversity conservation (Grumbine 1994). Since the early to mid-1990s, most federal agencies with mandates to conserve biodiversity have adopted policies for ecosystem management (Grumbine 1997). Two specific goals of ecosystem management are to maintain viable populations of all species and to ensure the preservation of all native ecosystems (National Ecosystem Management Forum 1993; Grumbine 1994). Because ecosystem management is also concerned with the conservation of biodiversity as well as with ecological relationships among species, ecological integrity is protected; ecosystem management, thus, might be viewed as natural systems management rather than natural resource management (Alpert 1995).

Ecosystem management has an advantage over single-species management because time and monies can be saved in the long term by implementing conservation measures for many coexisting species simultaneously instead of one or a few (Reed 1995). Furthermore, ecosystem management is attractive because it addresses a range of spatial scales from the forested tract to the landscape, recognizes that management may vary from one forested tract or landscape to another, appreciates that ecosystems are dynamic, and integrates ecological principles (Franklin 1997). The future success of ecosystem management, however, will be contingent on partnership and compromise among agencies (i.e., federal and state) and other stakeholders (e.g., private landowners) concerned with the conservation of biodiversity (Knight and Meffe 1997). A final challenge to successful ecosystem management will be the application of science to natural resource management; this ranges from applying concepts to actual field problems and situations, identifying and evaluating key properties of ecosystems, and establishing standards and protocols for inventorying, monitoring, and storing ecological databases (Christensen et al. 1996; Keddy and Drummond 1996; Reichman and Pulliam 1996).

A sustainable and well-managed forest ecosystem may be achieved with a compromise between the extremes of, on the one hand, preserving biodiversity regardless of costs (a "nature first" approach) and, on the other hand, exploiting resources in an ecosystem (a "people first" approach) (Gale and Cordray 1991; Salwasser 1991). Under certain circumstances, many species and biological communities may need special protection and be managed with the explicit goal of their long-term conservation (Robinson 1993). Thus, striking the best balance between ensuring the maintenance of environmental attributes, such as biodiversity, and the production of commodities, such as timber, will continue to be a major challenge to natural resource managers in the eastern deciduous forest (Frissell, Nawa, and Noss 1992; Slocombe 1993). Perhaps this balance can be achieved best on federal public lands in the United States, where a multiple-use approach is mandated by various federal agencies, such as the Forest Service and the Bureau of Land Management (Brussard, Murphy, and Noss 1992). In addition, various means of conserving biodiversity could be tested readily on these federal lands by developing management approaches to maintain or restore biodiversity while continuing to produce commodities for people (Thomas and Salwasser 1989).

We have little choice today but to address carefully and openly the issue of human demands for resources in the eastern deciduous forest from a sustainability and ecosystem-management perspective. Sustainability of biodiversity will lead to healthy ecological systems and wise resource management for recreational, economic, subsistence, and aesthetic purposes. Sustainability of biodiversity was one of three major research priorities proposed by the Ecological Society of America in its Sustainable Biosphere Initiative of 1988 (Lubchenco et al. 1991). Sustainability of the world's biodiversity, however, will be impeded by an ever-expanding human population, which is growing by 1.7 percent per year, a rate that translates into about 93 million more people added to our planet annually. Resources for this population and habitat for the world's biodiversity are finite—we need to confront this inescapable predicament. Moreover, maintenance and restoring biodiversity in all ecosystems worldwide as well as in the eastern deciduous forest will be essential to mitigate the impacts of future global climatic change, acid deposition, and other environmental concerns (Samson 1992). These concerns and their relation to biodiversity conservation in the eastern deciduous forest will be addressed in the following chapter.

## Conservation of Biodiversity and Forest Management

Forest management provides an opportunity to maintain long-term sustainability of resources and conservation of plant and animal diversity on private and public lands (Williams and Marcot 1991), which is vital to achieving no net loss in numbers of species (Brussard, Murphy, and Noss 1992). National and state parks and forests throughout the eastern United States are especially critical because these lands often contain relatively large forested tracts and are major sources of biodiversity (Yahner et al. 1994). Valley Forge National Historical Park and Gettysburg National Military Park in Pennsylvania, for instance, have comparatively large forested tracts with very diverse plant communities (Cypher et al. 1986; Yahner et al. 1992). At a statewide level, Pennsylvania and other states in the Northeast also are becoming increasingly aware of the need for the protection and management of forests for biodiversity conservation (Kirkland, Rhoads, and Kim 1990).

Management of the eastern deciduous forest in the future must be innovative and broad in scope. It must focus on preserving endangered and threatened species, and species of special concern; on restoring declining forests and preserving critical habitats; and on maintaining and restoring processes and functional relationships in forested ecosystems (McMinn 1991). Management of the forest at the ecosystem or landscape level, in particular, is the logical direction in the future to ensure biodiversity conservation (Brussard, Murphy, and Noss 1992; Oliver 1992; Grumbine 1994). Attention increasingly is being directed at this level to maintain biologically healthy and diverse habitats for forest plants and animals, resulting in a greater likelihood of preserving the biological diversity of species and their critical habitats (Radcliffe 1992; Franklin 1993). These management approaches will require cooperative interagency plans involving conservation biologists, wildlife managers, foresters, and other natural resource specialists. Unfortunately, cooperation among agencies, conservation groups, and others has not always come easily (Grumbine 1990).

The success of these management approaches will be contingent also on the involvement and cooperation of the many private forest landowners, who have a major stake in the conservation of biodiversity in the eastern deciduous forest. A high percentage of private ownership of forestlands will continue into the twenty-first century (Ticknor 1992). Thus, a real challenge will be to link wildlife managers, foresters, and private forest landowners in research and monitoring programs to promote biological conservation (Irwin and Wigley 1992). Private forest landowners could be instrumental by permitting their lands to be used to test the effects of forest-management practices on biodiversity. Replication of practices on several forest stands owned by private landowners will be necessary to address biological diversity concerns on the ecosystem or landscape scale. For this cooperation to succeed, landowners must first be convinced that natural resource managers recognize the economic value of forests. Success also requires the commitment of state and federal agencies (Irwin and Wigley 1992).

Forest management must be governed by a comprehensive policy incorporating science, economics, and politics to ensure that biological diversity concerns are met; this policy should contain three elements (Radcliffe 1992; see also Kuusipalo and Kangas 1994). First, forest management must have a long-term focus. Management of forests for the preservation of an endangered species while considering the economic ramifications (witness the northern spotted owl issue in the Pacific Northwest; Sample and LeMaster 1992) must move away from short-term, easy-fix solutions. Second, there must be a national commitment to establishing a set of priorities for managing public and private forestlands that creates a stable economic, political, and legal climate. Third, biodiversity conservation in forests must integrate biological concerns with a rational and careful understanding of the role and place of humans in healthy, productive, and sustainable ecosystems and landscapes and even in the biosphere.

Biodiversity conservation in the eastern deciduous forest on a statewide and even broader scale has been facilitated by a new methodology termed gap analysis (Scott et al. 1993). Already several eastern states, such as New York and Pennsylvania, are using gap analysis. This methodology identifies "gaps" in the representation of biodiversity in areas managed for the maintenance of native species or natural ecosystems. As gaps are recognized, new lands can be acquired or managed in such a manner as to conserve biological diversity. Priorities then can be established, for instance, to conserve endangered, threatened, and rare species or to emphasize the preservation of plant community types that existed in the landscape before European settlement (Strittholt and Boerner 1995; Kiester et al. 1996). In addition, gap analysis has raised the level of public awareness and debate on biodiversity conservation, thereby facilitating policy formulation and strategic planning for forested habitats and associated biota throughout the United States.

Another key element in successful biodiversity conservation in the eastern deciduous forest and throughout the United States is the establishment of the Biological Resources Division (BRD, formerly the National Biological Survey or Service) in 1996. The BRD was created to consolidate biological research, inventorying, and monitoring programs. This was the first national effort to streamline a dynamic inventory of plant and animal resources and their habitats under the Department of the Interior (Cohn 1993). The BRD gears research efforts to all levels of biodiversity, from endangered species to ecosystem management.

## Conservation Biology and Wildlife Management

Conservation biology emerged as a recognized scientific discipline in the 1980s specifically to focus on the application of science and biological principles, such as island biogeography and metapopulation theories, to the preservation of biological diversity (The Wilderness Society 1986; With 1997). Conservation biologists include a diverse group of scientists, ranging from population biologists and geneticists in biology, zoology, and botany departments to scientists in more traditional natural resource programs, such as wildlife management and forestry (Beissinger 1990).

Conservation biologists clearly share many common interests and approaches with wildlife and forest managers (Temple et al. 1988). All are concerned about the consequences of human activities on the environment and are interested and trained in the natural history of plants and animals. But a fundamental difference between conservation biologists and some scientists in more traditional natural resource programs is that the latter group historically dealt with the management of resources for economic or recreational purposes. For instance, wildlife professionals in the eastern United States have always been concerned with the management of white-tailed deer populations for recreational reasons. Only since the 1970s has the wildlife profession given attention to the management of many nongame species, such as songbirds.

The field of conservation biology perhaps originated as a direct consequence of the inability of natural resource management programs to consider and manage for all species, rather than just the economically important species in an ecosystem (Aplet, Laven, and Fiedler 1992). As early as 1949, Aldo Leopold encouraged natural resource managers to view their role as stewards of all species in the ecosystem rather than of a few (Leopold 1949). Fortunately, today many natural resource managers call themselves conservation biologists and are concerned about all species, not just those that are of economic value (Yahner 1990). Moreover, conservation biology can work in conjunction with wildlife management and forestry to help prevent extinctions of species and to maintain or restore biological diversity in sustainable ecosystems.

Professional societies will continue to play an active role in the conservation of biodiversity (Bawa and Wilkes 1992). The Society of Conservation Biologists, The Wildlife Society, and the Ecological Society of America are examples of professional groups that have taken a stance on this and other important issues (e.g., The Wildlife Society 1992). Some believe that professional societies have acted as a catalyst to mitigate the impacts of the United States' decision not to sign the biodiversity convention at the 1992 Earth Summit in Rio de Janeiro (Bawa and Wilkes 1992).

In summary, biodiversity conservation is a relatively recent but extremely timely issue in the minds of many people, including natural resource managers, scientists, and members of society in general (Yahner 1993b). Conservation of the diversity in the eastern deciduous forest, from genetic to landscape levels, will pose serious challenges to agencies and scientists from a variety of disciplines well into the next century. We are beginning to meet these challenges and already have some understanding of the impact of certain factors, for example, forest fragmentation, on forest biodiversity. Other factors, however, such as global climatic change, may have serious consequences for the future conservation of resources in the eastern deciduous forest. These factors and their effects on forest resources will be presented in the next chapter.

# 9. ATMOSPHERIC ENVIRONMENTAL CONCERNS

We have seen in earlier chapters that the distribution and abundance of plants and animals in today's eastern deciduous forest are affected by a spectrum of abiotic (e.g., solar radiation, soil moisture), biotic (e.g., predation, competition), and human-induced factors (e.g., forest-management practices, forest fragmentation). Since the 1960s, another set of environmental factors associated with the earth's atmosphere has become a major concern to scientists and the general public. These include the greenhouse effect, global climatic change, acid deposition, mercury deposition, tropospheric ozone buildup, and stratospheric ozone depletion. In this chapter, I focus primarily on the potential and real impacts of these factors on the ecology of and wildlife conservation in the eastern deciduous forest.

## The Earth's Atmosphere: Weather and Climate

Our earth is surrounded by three layers of gases, which collectively are called the atmosphere. The inner layer, or troposphere, extends from the earth's surface to about 17 kilometers into the atmosphere (Figure 9.1). It contains most (95 percent) of the earth's atmospheric gases, consisting mainly of nitrogen (78 percent) and oxygen (21 percent). Carbon dioxide (0.035 percent) and water vapor (0.01–5 percent, depending on latitude) constitute only a minute proportion of the tropospheric gases.

The stratosphere is the next layer of gases, extending 17 to 48 kilometers above the earth. Gases are much less dense in this middle layer. An important gas in the stratosphere is ozone, which is most abundant

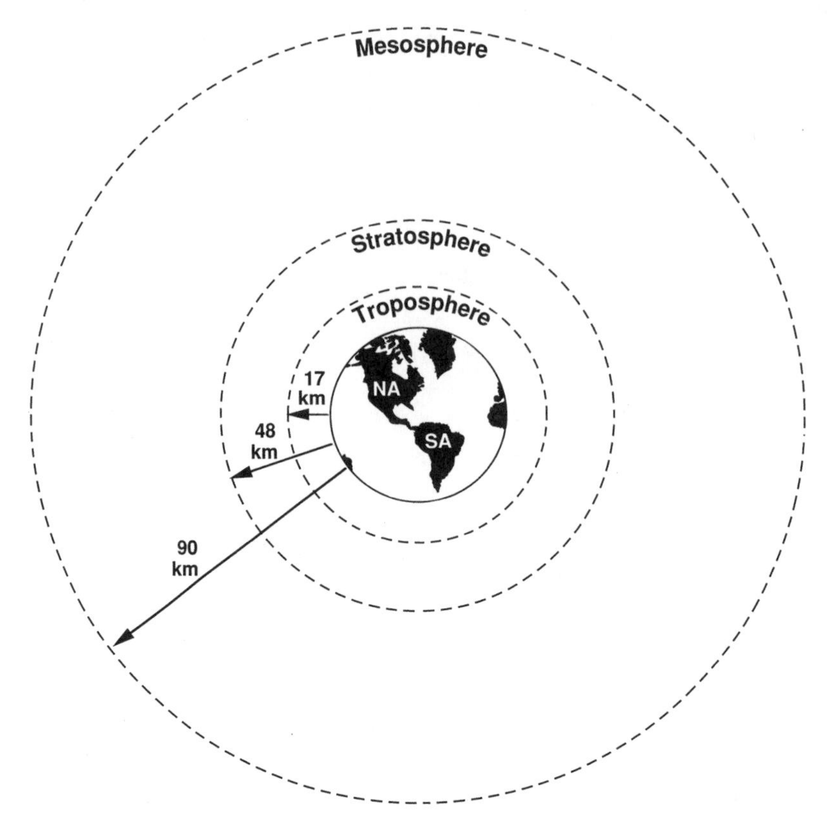


Figure 9.1. The three layers of gases surrounding the earth. Note that the earth is not drawn to scale; it has a diameter of 12,682 kilometers.

nearest the trophosphere, that is, between 17 and 26 kilometers above the earth's surface. Beyond the boundaries of the stratosphere and extending perhaps more than 90 kilometers is a third gaseous layer called the mesosphere.

Weather, defined as short-term changes in temperature, precipitation, wind speed, and other atmospheric conditions, takes place largely in the troposphere. Climate is the characteristic weather patterns of a given area and is the average temperature, precipitation, and other meteorological conditions expected to occur based on records kept over a period spanning at least a few decades (Easterling 1990). Climate in the eastern deciduous forest has changed relatively little over the past several thousand years following the retreat of the last glacial period (see Figure 2.1). This long-term stability in climate enables plants and animals to evolve adaptations to weather patterns, which in turn influence their current geographic distribution and abundance in the eastern deciduous forest.

Weather patterns can change unpredictably on occasion, resulting in droughts, cold waves, or heat waves. Perhaps the most dramatic weather event on earth is the occurrence of El Niño, which means "The Child," referencing to the Christ child (Ola and D'Aulaire 1997; Webster and Palmer 1997). This event was named in the nineteenth century by Peruvian sailors, who observed major warming and shifting of water currents every few years around Christmas. With an El Niño, warm water flows from the Pacific Ocean eastward toward North and South America rather than westward toward Asia. As a result, torrential rains, like those seen during 1997–98 along the Pacific coast of southwestern United States, Peru, and Ecuador, can occur with devastating regional consequences on people, economy, and wildlife; conversely, other parts of the world, such as Indonesia, New Guinea, and Australia, can simultaneously experience severe drought.

Regional changes in weather patterns may often have direct effects on the distribution and abundance of forest wildlife. For example, the drought of 1988 in Pennsylvania had a negative impact on small forest mammals, such as shrews (Yahner 1992). Natural weather phenomena, such as hurricanes, also can disrupt entire ecosystems. Hurricane Andrew in 1992 caused extensive damage to slash pine and the rare buccaneer palm in parts of southern Florida; fortunately, populations of valuable wildlife species, such as the endangered Florida panther, Cape Sable seaside sparrow, and red-cockaded woodpecker, were unaffected (Ogden 1992; Pimm et al. 1994).

If humans continue to dump unprecedented amounts of pollution into the earth's atmosphere, the climate and associated weather patterns may be altered in subtle but perhaps lasting manners. Because the atmosphere is vital to all organisms in the eastern deciduous forest, conservation of the atmosphere is in the long-term interest of natural resource managers and society in general. Collectively, the atmosphere is a common resource that is owned by no one, but it can be used and abused by anyone (Easterling 1990).

## Factors Affecting Climate

Temperature and precipitation are the major components of climate that influence the eastern deciduous forest and its resources. Average temperature and precipitation are determined mainly by the earth's radiation budget, which is the difference between the amount of the incoming energy from the sun (solar radiation) and the amount radiated back and beyond the reaches of the atmosphere (Easterling 1990). Approximately 50 percent of the radiation entering the mesosphere from the sun is absorbed at the earth's surface and by its vegetation. Another 20 percent is absorbed by radiatively active trace gases, such as water vapor, carbon dioxide, and ozone; these gases then help warm the atmosphere near the earth's surface. The remaining radiation (about 30 percent) is bounced back or scattered to space by clouds, gases in the atmosphere, or the earth as infrared, or long-wave, radiation.

### *Seasons and Earth-Sun Relationships*

A change in the earth's radiation budget is the driving force behind predictable climatic events, such as seasons, in the eastern deciduous forest. Early growth of deciduous foliage in the early spring and dieback of leaves in autumn occur annually as the amount and timing of incoming solar radiation differ as a function of two earth-sun relationships: the distance between the earth and sun and the tilt of the earth on its axis (Easterling 1990; Miller 1990). Distances from the earth to the sun vary seasonally because of the shape of the earth's annual orbit. In summer and winter, the earth is farthest from the sun. Furthermore, the north-south axis of the earth is tilted throughout the year at an angle of 23.5 degrees with respect to an imaginary plane running through the center of the earth and the sun (Easterling 1990). This tilt of the earth allows the rays of the sun to strike the eastern deciduous forest (as elsewhere in the northern hemisphere) most directly in summer and least in winter.

The tilt of the earth on its axis also causes movement of large air masses in the troposphere (Miller 1990). Air masses near the equator become warm with the influx of solar radiation; these warm masses then rise and spread northward and southward toward the poles, thereby moderating summer temperatures in the eastern deciduous forest. The earth's tilt combined with its daily rotation on its axis can affect the movement of large air masses in the troposphere and deflect these in a westward or an eastward pattern as prevailing surface winds; the prevailing surface winds found in the eastern United States are the west trade winds. As these trade winds flow easterly, moisture is picked up and released as precipitation, which is vital to plant species characteristic of the eastern deciduous forest. In addition, the topography of an area can influence the movement of prevailing surface winds (Miller 1990). When prevailing winds flow toward a mountain ridge, the winds are deflected upward, causing the air to cool and lose precipitation on the windward side. The opposite slope, or leeward side, is therefore traversed by dryer air, a phenomenon termed a rain shadow effect. This phenomenon is more prevalent in western North America, where mountains are of higher elevations than those in eastern North America.

### *Greenhouse Gases and the Greenhouse Effect*

Absorption of solar radiation by both the earth and radiatively active trace gases, often referred to as greenhouse gases, results in a heat buildup in the troposphere called the greenhouse effect (Laarman and Sedjo 1992). Chlorofluorocarbons are gases derived from methane or ethane in which hydrogen atoms are replaced with chlorine and fluorine. Three important greenhouse gases that occur naturally in the troposphere are carbon dioxide, water vapor, and ozone. High-energy, visible light from the sun passes through the earth's atmosphere but is reflected back as low-energy, infrared radiation. Carbon dioxide and other greenhouse gases trap this low-energy radiation as heat, which warms and insulates the earth in a manner similar to a pane of glass in a greenhouse. Ozone in the stratosphere is especially relevant to life on earth because it traps about 99 percent of the harmful ultraviolet (UV), or short-wave, radiation before it enters the troposphere. By trapping harmful UV radiation, ozone also warms the air in the stratosphere well above the earth's surface (Miller 1990). In contrast to ozone, aerosol gases, introduced into the upper atmosphere by certain types of volcanoes, generally reflect incoming solar radiation and cause cooling of the earth and its atmosphere (Easterling 1990).

Without a natural greenhouse effect in both the troposphere and the stratosphere, life on earth as we know it would not be possible (Easterling 1990). Temperatures on our planet average 13°C, which is 33°C warmer than it would be without the greenhouse effect (World Resources Institute 1990). By comparison, Venus has excessive amounts of atmospheric carbon dioxide, which produce a pronounced greenhouse effect and average temperatures of 447°C. Mars, on the other hand, has virtually no atmospheric gases and has average temperatures of -53°C. Both of these planets are devoid of life, which suggests that any changes in levels of greenhouse gases could have detrimental impacts on life on earth.

## Global Climatic Change

Since the 1970s, the scientific community has begun to agree that global climatic change is inevitable (Cinq-Mars, Diamond, and LaRoe 1991; Root and Schneider 1993). If fact, global climatic change has probably been occurring for the past 150 years, beginning perhaps with the Industrial Revolution and subsequent expansion of the world's economy (Easterling 1990). Over this period, levels of the greenhouse gases carbon dioxide and methane have increased 25 percent and 100 percent, respectively. This increase has translated into about a 0.5°C increase in global temperatures in the twentieth century mainly because of additional infrared radiation being trapped as heat by these gases in our atmosphere. Major droughts, powerful hurricanes, destructive forest fires, and substantial flooding in the late 1980s brought widespread public attention to the prospect of pending global climatic change (Fajer 1989; Easterling 1990; Office of Technology Assessment 1993).

Although the consequences of global warming on the earth's resources are still uncertain, we are now beginning to appreciate that future climatic change will probably have several measurable impacts on all ecosystems of the world, including the eastern deciduous forest. Some likely effects of global warming will be a longer growing season, allowing the expansion of a grain belt farther northward into Canada; a change in wind and ocean currents; and a rise in sea level of 0.3 to 1 meter because of melting mountain glaciers, reduction in the extent and thickness of Antarctic and Arctic ice sheets, and expansion of sea water as the ocean temperatures rise (World Resources Institute 1990; Ojima, Galvin, and Turner 1994; Barron 1995). If sea levels rise 1 meter because of global warming, about 3 percent of the world's land mass will be flooded (equivalent to an area the size of the United States west of the Mississippi River), which is home to more than one billion humans (Myers 1993). Equally alarming is the fact that if global warming occurs, it is expected to cause shifts in the distribution and perhaps even the extinction of an unknown number of fauna and flora species (World Resources Institute 1990).

### *Recent Trends in Climate in the United States*

Global warming in the twentieth century has generally been more pronounced in the Southern than in the Northern Hemisphere (Easterling 1990). In the United States, summer temperatures gradually have increased since the 1970s, but mean temperatures in winter, spring, and autumn have tended to be cooler. Concurrently, a shift toward increased precipitation has taken place, and droughts that have occurred since the 1970s have been less severe than those in the 1930s and 1950s (Easterling 1990).

### *Predictions in Global Climatic Trends*

The link between higher atmospheric levels of greenhouse gases, such as carbon dioxide, and increased global warming is well documented (e.g., World Resources Institute 1990; see also Wheeler 1990). Thus, the debate today is not about whether global change will occur but rather about its magnitude and rate. The recent rise in global temperature of 0.5°C is believed to have occurred about ten times faster than the rise during the last interglacial period (Cinq-Mars, Diamond, and LaRoe 1991).

A total of 241 billion metric tons of carbon dioxide has been released by human activities from 1860 to 1987. Of this total, consumption of fossil fuels accounted for 181 billion metric tons (75 percent), and deforestation contributed the remaining 60 metric tons (25 percent) (World Resources Institute 1990). Most fossil-fuel consumption has occurred since 1950: 130 million metric tons, compared to 51 million metric tons from 1860 to 1949. In 1987, for example, 1.2 million metric tons of carbon dioxide were emitted into the atmosphere by the United States alone, while the same amount was given off as a result of intensive land clearing by fire in the Amazon basin of Brazil. By far, the leading producer of atmospheric carbon dioxide from 1950 to 1987 was the United States, followed by the European Community and the former Soviet Union.

The total release of 241 billion metric tons of carbon dioxide from 1860 to 1987 is significant especially when we consider that the world's forests can store only about 450 billion metric tons of carbon; this capacity to keep carbon from entering the atmosphere is lost once the forests are cleared. Fortunately, regrowth of forests, such as the continual maturation of the forest in the eastern United States, may help increase carbon storage and offset carbon loss due to the greenhouse effect (Sedjo 1990). Large, healthy trees not only are important for storing carbon and mitigating high concentrations of carbon dioxide in the atmosphere but also improve air quality by removing harmful air pollutants, such as sulfur dioxide (Nowak 1994a, 1994b). By the end of the twentieth century, most (67–80 percent) of the increase in atmospheric carbon dioxide will be caused by the burning of fossil fuels, such as coal and gasoline, with the remainder caused by clearing and burning of tropical forests (World Resources Institute 1990; Laarman and Sedjo 1992).

Today, considerable alarm exists about increased levels of other gases, most notably chlorofluorocarbons (CFCs) and methane, which can substantially increase the greenhouse effect (Laarman and Sedjo 1992). For instance, compared to the heat captured by one molecule of carbon dioxide, twenty thousand times as much heat is captured by one molecule of CFC, and twenty to thirty times as much heat is trapped by one molecule of methane (World Resources Institute 1990; Neue 1993). As a result, the global temperature is expected to increase between 0.5 and 4.5°C by the first few decades of the twenty-first century unless atmospheric levels of greenhouse gases are curbed. By the year 2020, carbon dioxide is expected to cause 50 percent, CFCs 25 percent, methane 15 percent, and nitrous oxide 10 percent of this increase in global warming. Burning of fossil fuels and deforestation will contribute to carbon dioxide buildup; leaking air conditioners and refrigerators and chemicals used in aerosol cans will be responsible for CFCs; decay of bacteria in rice patties and wetlands and leaks from natural gas wells, pipelines, and other sources will cause methane to rise; and fertilizers and livestock waste will contribute to nitrogen oxide increases (Laarman and Sedjo 1992; Bubier, Moore, and Roulet 1993; Neue 1993).

Another plausible effect of global climatic change is greater climatic variability and severity (Easterling 1990; Barron 1995). Heat waves and severe storms may continue to become more frequent, whereas cold periods and damaging frosts may occur less often. Hurricanes of the twenty-first century, for instance, may hit the coastline of the eastern United States farther north, occur during more months per year, and be 50 percent more powerful than those of the twentieth century (Miller 1990). This greater climatic variability and severity will probably have major, but as yet unknown, impacts on the distribution and the abundance of forest flora and fauna.

### *Global Circulation and Regional Climate Models*

Estimates of a 0.5 to 4.5°C rise in global temperature in the early decades of the twenty-first century are based on predictions obtained from large and complex global circulation models (GCMs) or regional climate models (RegCMs) (Easterling 1990; Barron 1995; Bartlein,

Whitlock, and Shafer 1997). These models are mathematical attempts to simulate future, large-scale climatic changes by using daily weather information, but they have several inherent weaknesses. First, because the climate is very complex, it cannot be simulated completely using existing GCMs or RegCMs (Cooter et al. 1993; Root and Schneider 1993; Bartlein, Whitlock, and Shafer 1997). Second, the resolution of these models is too crude to separate regional changes in climate on a scale that is meaningful to natural resource managers in the eastern deciduous forest. For example, climatic changes at the ecosystem level, e.g., a 50-by-50-kilometer or smaller grid, would be very useful to a manager; however, most GMCs, for instance, have used a much larger scale, such as a grid 4.5° latitude by 7.5° longitude. Third, the climatic variables important to forest and wildlife managers for an understanding of potential consequences of global climatic change on plants and animals, such as soil moisture, are rarely used in today's GCMs or RegCMs. Fourth, a prediction of weather patterns for the next year based on one year of weather information in the 4.5°-by-7.5° grid would take only several hours of computer time using technology available in the early 1990s, whereas the same prediction using the 50-by-50-kilometer grid would require a full year of computer time (Root and Schneider 1993). In short, GCMs and RegCMs are impractical and too costly to assess climatic change at the ecosystem level; they currently do not have the capacity to examine the effects of future climatic patterns on plants and animals associated with a given forest cover type or land use in the eastern deciduous forest.

## Predicted Impacts of Global Climatic Change on Biota

On a worldwide basis, boreal forest and tundra in North America are considered to be most susceptible to future climatic warming; some scientists speculate that the rate of global climatic warming in more northerly latitudes will occur at twice the rate of warming in more southerly latitudes (Mather 1990; Bonan, Pollard, and Thompson 1992). Hence, plants and animals adapted to boreal forest and tundra will be more at risk than those in the eastern deciduous forest (World Resources Institute 1990). Other regions of North America that are expected to be affected dramatically and rapidly by global warming are the coasts. As sea levels rise, plants and animals of barrier islands, freshwater estuaries, and coastlines, for example, migratory shorebirds, may be affected adversely (Figure 9.2).

Figure 9.2. The laughing gull is a bird of coastal areas. (Photograph by Kurt G. Engstrom, School of Forest Resources, Pennsylvania State University.)

### *Impacts on Plants*

As discussed in chapter 3, trees and other plant life have evolved adaptations to various ecological conditions, such as temperature and soil moisture. Current patterns of the distribution and abundance of forest plants are mainly a function of a long-term response to these conditions, which may have taken hundreds or even thousands of years to evolve. A rise of a few degrees in global temperature over the next century, however, could force changes in the geographic ranges of plants. Populations of some trees, such as sugar maple, may be able to expand their range northward while possibly experiencing dieback of populations in warmer southerly latitudes (Figure 9.3) (Root and Schneider 1993). Other trees species in the eastern deciduous forest, such as beech, birch, and hemlock, would have to shift geographic ranges at least 500 kilometers northward to disperse fast enough to keep track with northerly shifts in warmer climatic conditions (Cohn 1989; Office of Technology Assessment 1993). Unfortunately, range shifts by most plant species to more northerly latitudes probably would not keep pace with a steady warming trend, because plant species can usually disperse only about 10 to 40 kilometers per century (Dobson 1992). Also, species with smaller geographic ranges would be more likely to go extinct than those with larger ranges (Peters 1988). Furthermore, seedlings will be more vulner-

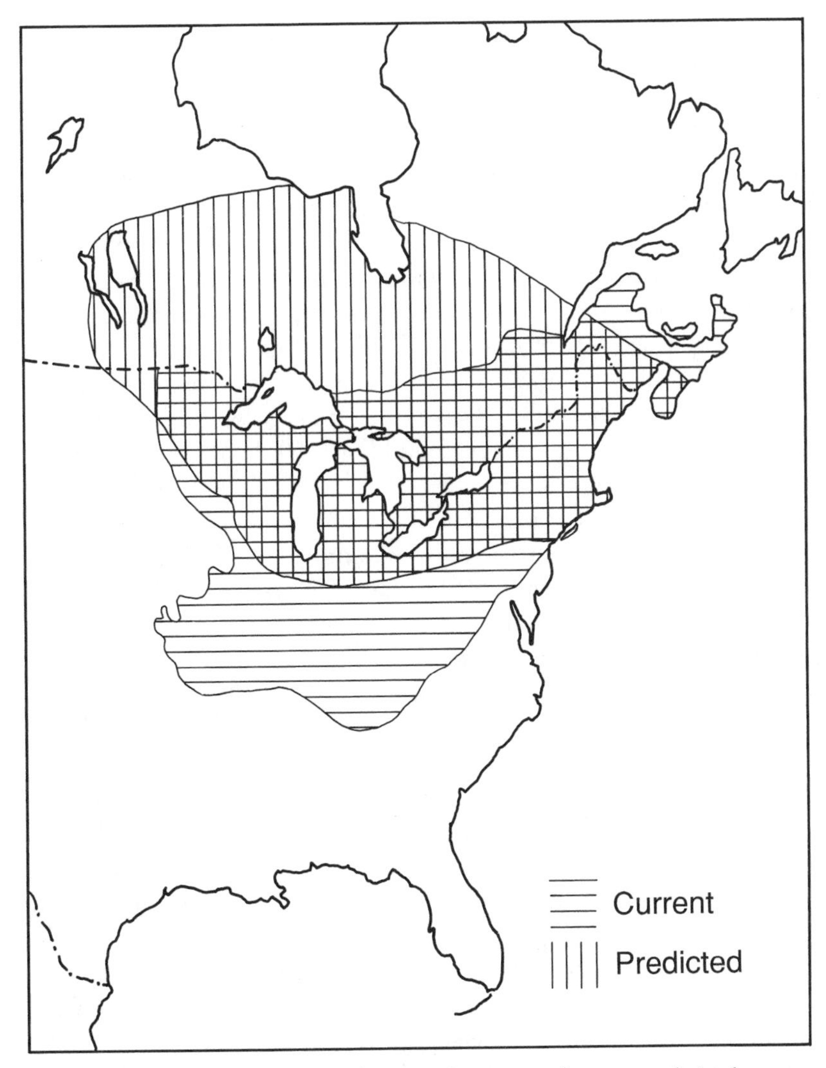


Figure 9.3. Current and predicted geographic range of sugar maple in the eastern deciduous forest as a result of a global warming trend. (Modified from Root and Schneider 1993.)

able than mature trees to a climatic change. As these seedlings are lost, changes in the species composition of a forest could occur because mature trees that eventually die would not be replaced. Hence, the end result of global warming may be massive extinctions of plants in the eastern deciduous forest at a rate unprecedented in history. In some cases, plant communities would become seriously disrupted with the loss or recent addition of species, which would have unknown implications for the ecology of the eastern deciduous forest (e.g., see Root and Schneider 1993).

The frequency and intensity of forest fires are likely to increase with future global warming (Franklin et al. 1991). Outbreaks of insect pests and diseases also may be expected if forested tracts are stressed by changes in climate. If global warming affects long-term timber yields, as predicted by some scientists (see World Resources Institute 1990), forest managers will be forced to manage the eastern deciduous forest differently (Hagenstein 1990). As a hypothetical example, if global warming were to become pronounced in the eastern deciduous forest, clear-cutting might no longer be a viable forest-management option because of increased temperatures near ground level and the resultant stress on germinating seedlings. Time still remains to begin an evaluation of various forest-management practices that are both practical and environmentally sound in order to help mitigate the potential effects of global warming on the forest (Office of Technology Assessment 1993).

The projected changes in climate combined with their potential repercussions on vegetation present important challenges to current management and conservation goals (Bartlein, Whitlock, and Shafer 1997). At least five options are available to natural resource managers to minimize the negative consequences of any future global warming for the eastern deciduous forest. First, tree-planting programs are important because trees help remove carbon dioxide from the earth's atmosphere (Kurz and Sampson 1991). To offset the carbon dioxide emitted from the global burning of fossil fuels, however, would require an addition of 1 billion hectares of forests in the next forty to fifty years (World Resources Institute 1990); this amount represents about 25 percent of the total forest area currently present in the world (World Resources Institute 1988). Second, a diversity of tree species obtained from non-local seed sources should be planted (Ledig and Kitzmiller 1992). These seed sources could be imported from populations in more southerly latitudes or at lower elevations so that they would produce trees that are better adapted to a warmer climate. Third, tree-improvement programs should begin testing seed sources of deciduous trees that demonstrate adaptability to warmer climates. Fourth, forest tree seeds should be conserved to offset the effects of a rapid climatic change on the genetic diversity of the forests throughout the United States (Office of Technology Assessment 1993). Finally, connectivity in the landscape caused by creating and preserving habitat corridors may provide pathways for dispersal or movement by plants in response to warming conditions (Bartlein, Whitlock, and Shafer 1997).

### *Impacts on Animals*

Global warming could have both negative and positive impacts on forest animals. Progressively warmer winters in the eastern deciduous forest could lead to dramatic spring increases in the abundance of insect populations, such as caterpillars, that can extensively defoliate forest trees and, hence, be a negative effect (Carroll 1992). Thus, a once healthy forest stand that is already stressed by extreme climatic change could become even more susceptible to insect or disease damage (World Resources Institute 1990). Conversely, some researchers contend that milder winters lower the abundance of forest insects by increasing the abundance of predators and parasites that rely on these insects as food or hosts. As a result, breeding populations of warblers, for example, the black-throated blue warbler of the northern hardwood region, may increase if warmer winter temperatures result in higher insect abundance in spring (Rodenhouse 1992).

Another good example of a bird whose distribution and abundance may be benefited by global warming is the eastern phoebe. The winter range of the phoebe is restricted by January temperatures (Root and Schneider 1993) (Figure 9.4). As global climatic change warms the more northerly latitudes in the eastern deciduous forest over the next several decades, the range of the phoebe is expected to shift northward. Conversely, as winters continue to warm, occurrences of some birds, such as the golden-crowned kinglet (Yahner 1993a), may become more common in more southerly latitudes of the eastern deciduous forest in the winter months. In fact, at least fifty other species of songbirds throughout North America have northern boundaries of their wintering ranges that are limited by low temperatures; conceivably, global warming will influence the extension of the wintering ranges of many of these species (Root 1988).

Small mammals, which have limited abilities to expand their geographic range compared to more mobile birds, may be relatively more susceptible to global climatic change (Cohn 1989). If global warming changes the habitat used by small mammals, local populations may be extirpated. For instance, global warming of 3°C is projected to cause a loss of 9 to 62 percent of the total small mammal species on nineteen isolated mountain ranges in the Great Basin of the western United

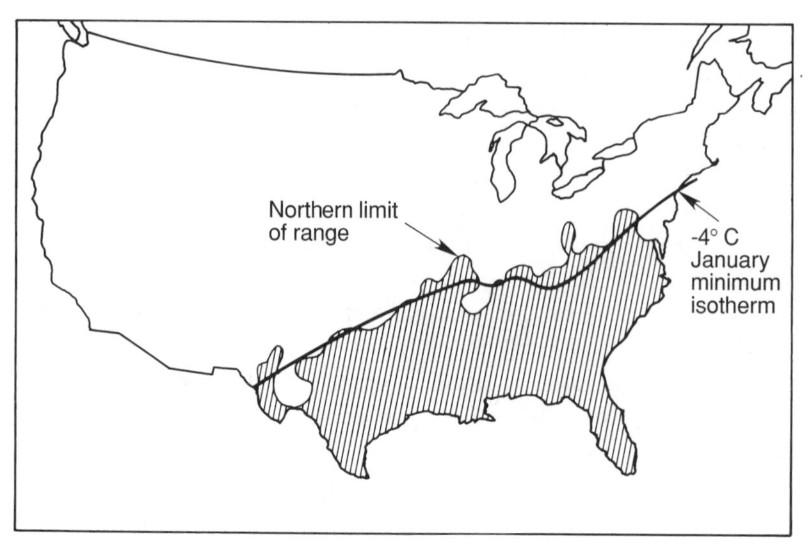


Figure 9.4. Winter range distribution of the eastern phoebe in North America. (Modified from Root and Schneider 1993.)

States (McDonald and Brown 1992). In Texas, changes in microclimatic conditions associated with caves and cavities used as hibernacula by bats are expected to become warmer, thereby having an adverse effect on bat species richness in the future (Scheel, Vincent, and Cameron 1996). From another perspective, drought conditions induced by global warming may result in greater availability of cracks in the soil, which in turn would be beneficial to small mammals using these cracks as refugia from predators (Stokes and Slade 1994). Although comparable predictions of changes in geographic distribution or microclimatic conditions are not yet available for mammals in the eastern deciduous forest, more northerly species of small mammals that occur in isolated pockets along the Appalachian Mountains (e.g., northern flying squirrel, rock vole, and New England cottontail) could be affected negatively by global warming.

Some species of mammals in the northeastern United States and southeastern Canada, such as Appalachian woodrats, eastern cottontails, and white-tailed deer, have northerly geographic ranges that are limited by winter climatic conditions. Therefore, we can speculate that global warming may facilitate the expansion of these species farther north in coming decades. Just the opposite trend has been documented in several mammal species, probably in response to lower than expected annual temperatures in the central plains of the United States from the 1930s through the 1980s. Over these six decades, four northerly species of small mammals, including masked shrews, meadow jumping mice, meadow voles, and least weasels, have expanded their current geographic ranges southward (Frey 1992) (Figure 9.5).

The distribution and abundance of certain wildlife species may be influenced indirectly by the impacts of global warming on certain features of their habitat. For example, the red-cockaded woodpecker of the Southeast depends on mature living trees for nesting. If global warming causes high mortality of mature trees in the breeding range of this woodpecker, localized populations could be extirpated (Root and Schneider 1993). In northern Michigan, the few remaining Kirtland's warblers depend on jack pine as nesting habitat (see Figure 7.2); their ground-level nests are placed beneath jack pines on sandy, well-drained soil (Botkin, Woodby, and Nisbet 1991). A modest climatic change could force jack pines to disperse northward to areas with less sandy soil where poorer drainage may occur. Survival of nestling Kirtland's warblers, therefore, would be reduced, probably leading to the extinction of this warbler within thirty to sixty years (Root and Schneider 1993).

Aquatic vertebrates also are at risk with the onset of global warming.

Figure 9.5. The least weasel is the smallest carnivore in the world. (Photograph by D. Pattie.)

For instance, the productivity of fishes adapted to cold waters during summer, such as trout, may decline as global warming increases the temperatures of streams and lakes (Mlot 1989; Cinq-Mars, Diamond, and LaRoe 1991). On the other hand, fishes adapted to warmer waters may be at an advantage; productivity of walleyes in Lake Michigan is predicted to increase 29 to 33 percent because of global warming (Mlot 1989). The distribution and the abundance of cold-intolerant animals, such as amphibians and reptiles, may be enhanced by global warming (Cinq-Mars, Diamond, and LaRoe 1991).

In general, we can expect global climatic change to be yet another factor that will magnify the rates of biodiversity loss worldwide (Wilson 1989; Dobson 1992), especially in mature and old-growth forests (Cinq-Mars, Diamond, and LaRoe 1991). Biodiversity will suffer much greater impacts if the globe becomes warmer and dryer rather than warmer and wetter. Impacts will be greater because the vast storehouse of the world's species occurs in the warm, moist latitudes containing the tropical forests and a tremendous biodiversity of plants and animals (World Resources Institute 1990). Clearly, nations will need to prepare global plans to deal with the pending impacts of global climatic change on the world's biota (Flavin and Lenssen 1994). In 1997, the international community took some initial steps by meeting in Kyoto, Japan, with the intent to begin formulating a global climate treaty (O'Meara 1997). At this meeting, 160 nations developed a protocol to reduce the world's emissions of greenhouse gases. Likewise, natural resource managers in the eastern deciduous forest should begin to prepare for the potential effects of global climatic change on its forest resources.

## Acid Deposition

Sulfur dioxide and nitric oxide are two major types of air pollutants in the troposphere that result principally from human activities; these pollutants are called primary air pollutants (Miller 1990). In the latter part of the twentieth century, 90 to 95 percent of the sulfur dioxide and 57 percent of the nitric oxide in the United States have been emitted by electric power plants and factories that burn fossil fuels, such as coal and oil. Automobile emissions are another major source of nitric oxide to the earth's troposphere. When transported by winds in the troposphere, both sulfur dioxide and nitric oxide react chemically to form new chemicals, such as sulfuric acid and nitric acid. These chemicals, called secondary air pollutants, fall to the earth's surface with rain or snow as acid deposition (sometimes referred to as acid precipitation or acid rain). Since World War II, the pH of precipitation in the eastern United States has decreased from about 5.6 to near 4.0; this greater acidity is attributed mainly to increased levels of sulfuric and nitric acids in the atmosphere (Galloway et al. 1982; Chappelka 1987). Thus, acid deposition today can be at least twenty to thirty-five times more concentrated than uncontaminated, natural precipitation (Lynch et al. 1993), such as that found before the Industrial Revolution.

The repercussions of acid deposition are varied. It may aggravate or even cause respiratory diseases in humans, degrade water quality by leaching toxic chemicals, and affect reproduction or increase mortality in plants and animals (Miller 1990). Acid deposition also has damaged cultural resources, ranging from historic buildings and outdoor sculptures in Montreal to historic monuments at Gettysburg National Historic Park (Sherwood and Dolske 1991; Weaver 1991). The effects of acid deposition on the eastern deciduous forest are not completely understood. However, damage caused by this deposition and costs to reduce emissions of primary pollutants in the United States alone will possibly run into the billions of dollars (Miller 1990).

### *Extent of the Acid Deposition Problem*

Acid deposition was first described as a problem in Europe in the early nineteenth century and has been studied extensively in North America since the 1970s (Blancher 1991). It is particularly a problem in certain parts of the world, including the northeastern United States, southeastern Canada, northern and central Europe, China, the former Soviet Union, and South America (Schindler 1988; Miller 1990). In the United States, about 90 percent of the emissions of sulfur dioxide and nitric oxide are produced in the thirty-one states east of the Mississippi River (Nash, Davis, and Skelly 1992). Emissions from industrialized states, such as Indiana, Illinois, Missouri, Ohio, Pennsylvania, and Tennessee, have resulted in acid deposition with a pH near 4.0 (Figure 9.6). These emissions are spread by winds over much of the northeastern United States and across the border into southeastern Canada.

The amount of acid deposition can vary dramatically over relatively short distances, in part because of differences in amounts of precipitation. In Pennsylvania, for instance, precipitation was greater in the western part than in the eastern part of the state during 1992; consequently, acid deposition in 1992 was significantly higher in the western (average

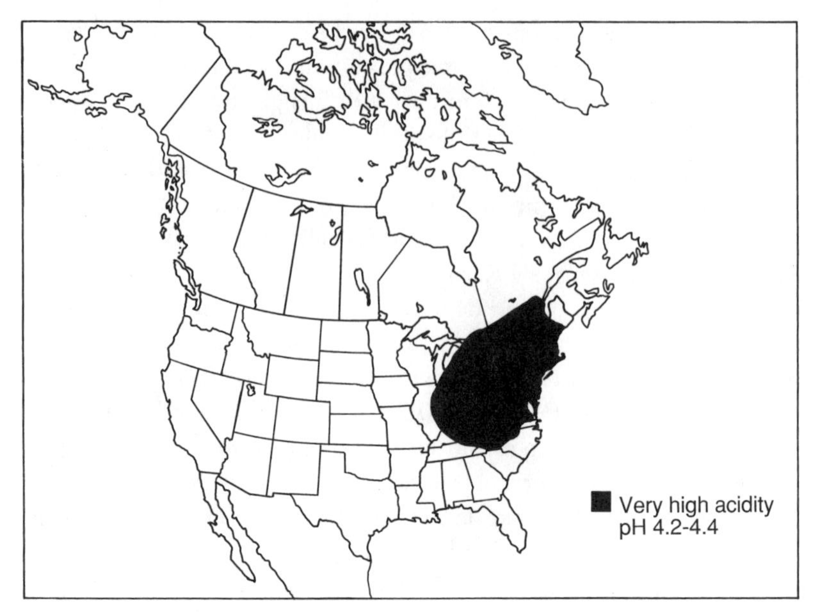


Figure 9.6. Map of the eastern half of North America showing the area in northeastern United States and southeastern Canada with precipitation that ranges in pH from 4.2 to 4.4. (Modified from Miller 1990.)

pH = 4.17) compared to the eastern part of the state (average pH = 4.30) (Lynch et al. 1993).

## *Impacts on Plants and Soils*

The impact of acid deposition on forest plants and soils emerged as a natural resource issue beginning in the early 1980s (Ruark et al. 1991; Laarman and Sedjo 1992). Sulfur dioxide has been known to be toxic to plants since the late nineteenth century (Skelly et al. 1989). Symptoms of sulfur dioxide injury to trees are irregularly shaped patches of tan-colored, dead tissue on surfaces of leaves, which can affect photosynthesis (Figure 9.7). The long-term consequences of acid deposition on forest ecosystems, however, is uncertain. Acid deposition generally has not produced "visible" effects on the health of the forest in the Northeast (de Steiguer, Pye, and Love 1990; U.S. Department of Agriculture 1992), but this issue is still debated in the scientific community. Some scientists (e.g., Nash, Davis, and Skelly 1992; Skelly 1989, 1993; Bennett et al. 1994) strongly advise against blaming acid deposition for forest declines when adequate data to support such "cause-and-effect" relationships do not yet exist.

Figure 9.7. Injury to oak leaves exposed to sulfur dioxide. (Photograph by Donald D. Davis, Department of Plant Pathology, Pennsylvania State University; from Skelly et al. 1989.)

Limited evidence suggests that acid deposition has caused extensive mortality of certain localized forest stands in the Northeast (Miller 1990). In particular, acid deposition is blamed in part for killing 60 percent of the red spruce trees in stands at high elevations in New Hampshire, New York, and Vermont. These spruce trees are potentially vulnerable because they are long-lived and exposed year-round to air currents carrying acid deposition and other pollutants. Yet despite the death of mature spruce trees in the Northeast, younger spruce trees are growing well in some of these stands (Erb 1987). A plausible cause-and-effect relationship, however, has been documented in the Bay of Fundy, New Brunswick, to explain injury to leaves of paper birch and mountain paper birch from acid marine fogs (Cox, Lemieux, and Lodin 1996).

Suspected tree die-offs resulting from acid deposition and other airborne pollutants are not restricted to the Northeast but have been reported throughout the Appalachians from southern Canada to Georgia (Miller 1990). In Europe, acid deposition and other airborne pollutants have been blamed, although with much controversy, for the destruction and overall decline in health of nearly 25 percent of the forests (Skelly and Innes 1994). Acid deposition, for instance, has been singled out as the factor causing a decline in health of silver fir and Norway spruce in

Europe (Ledig 1992). Other factors contributing to forest damage in North America, Europe, and perhaps elsewhere are disease, insect pests, ozone, and nutrient deficiency (Mather 1990; Laarman and Sedjo 1992; Skelly 1992a, 1993). Damage to forests by gypsy moth populations, for instance, is probably much more detrimental to the health of our forests than airborne pollutants (Skelly 1992b).

Long-term and increased soil acidity resulting from acid deposition is believed to affect nutrient cycling in forest soils by leaching chemicals, such as calcium and magnesium, that are important to the nutrition of forest trees and by killing valuable soil microorganisms that are critical to the normal functioning of a forest ecosystem (Miller 1990; Drohan and Sharpe 1997). These effects on forest soils are most dramatic in areas with shallow, acidic soils; in areas with limestone soils, which are quite basic, acid deposition may be neutralized (Mather 1990).

### *Impacts on Animals*

Airborne pollutants can have a long-term and dramatic impact on animals in the eastern deciduous forest. A notable case study is that of zinc deposition from two smelters along Blue Mountain in eastern Pennsylvania and its effect on wildlife distribution and abundance (Figure 9.8). Within 3 kilometers of the smelters, more than 800 hectares of the north-facing slope were nearly devoid of vegetation, and few species of vertebrates (amphibians, birds, and mammals) occurred within 5 kilometers east (downwind) of the smelters (Storm, Yahner, and Bellis 1993).

Acid deposition can also have negative consequences for animals. For instance, acid deposition significantly changes the chemistry of streams and lakes, which results in increased mortality, reduced growth rates, and lower reproductive success of fish (Mills and Schindler 1986). Increased acidification of waters can alter the species composition of fish communities by making water conditions unsuitable to acid-intolerant species and by destroying food resources of fishes. Furthermore, increased acidification releases metals (aluminum, mercury, and so on) in the soil, which leach into water, thereby affecting fish physiology by preventing respiration (Moyle and Leidy 1992).

The negative effects of acid deposition on aquatic organisms have been well studied over the past couple of decades. In Pennsylvania, populations of brook trout and aquatic insects, which are important food for trout, have been affected severely by acid deposition (Arnold et al. 1981; Sharpe 1990). In sixty-one headwater forest streams in southwestern

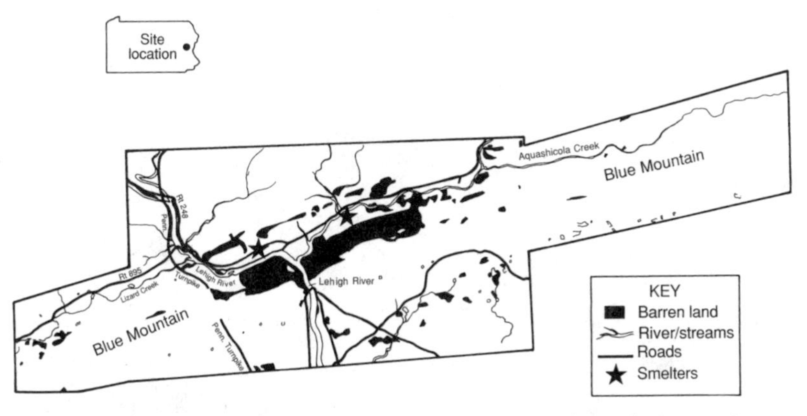


Figure 9.8. Barren areas (devoid of significant vegetation) in relation to zinc deposition from two smelters along Blue Mountain in eastern Pennsylvania. (Modified from Storm, Yahner, and Bellis 1993.)

Pennsylvania, only twenty-eight (46 percent) had viable populations of brook trout, because of low pH caused by acid deposition (Sharpe et al. 1987). As a result of acidic conditions, the Pennsylvania Fish and Boat Commission has removed sixteen streams or portions of streams in the state from its fish-stocking list since 1969, which amounts to about 1.7 percent of the total length of streams that are stocked annually with trout in Pennsylvania (D. E. Spotts, Pennsylvania Fish and Boat Commission, personal communication). In thirteen streams of the Adirondack Mountains, Catskill Mountains, and northern plateau in Pennsylvania, high concentrations of aluminum resulted in increased mortality of a variety of fish species, such as blacknose dace, slimy sculpin, and mottled sculpin (Van Sickle et al. 1996). Generally, these fish species were absent from small streams with both low pH (< 5.20) and high inorganic aluminum concentrations (> 100 µg/l) (Baker et al. 1996). In Ontario, the diversity of fish species declined 40 percent in lakes altered by acid deposition (Matuszek and Beggs 1988). In the Adirondack Park, New York, about 40 percent of the smaller lakes have become too acidic for survival of fish, regardless of species. Clearly, the biological integrity of many lakes and streams in the eastern deciduous forest will continue to be affected seriously if acid deposition does not decline.

Many amphibians, including salamanders, frogs, and toads, breed in temporary and permanent bodies of water (Pough and Wilson 1977). These breeding habitats often have low pH (below 5.0), presumably because of acid deposition, which can lower egg and larval survival in

Figure 9.9. A spotted salamander. (Photograph by Robert P. Brooks, School of Forest Resources, Pennsylvania State University.)

amphibians (Freda 1986; Freda, Sadinski, and Dunson 1991). In central Pennsylvania, for instance, no larval Jefferson salamanders survived to the adult stage in ponds with a pH of 4.2 (Sadinski and Dunson 1992). A coexisting woodland salamander, the spotted salamander (Figure 9.9), occasionally survived from larval to adult stage under the same acidic conditions, but adults that survived weighed less and metamorphosed more slowly into adulthood. Similarly, growth and development of frog tadpoles are affected in ponds with low pH (Cummins 1986; Sadinski and Dunson 1992). By comparison, acid deposition did not cause declines in a study of amphibians occurring in high-elevation and semi-temporary ponds of the Rocky Mountains (Vertucci and Corn 1996).

Amphibians also are susceptible to high acidic conditions in forest soils created by acid deposition. In New York, for example, forested habitats with relatively high soil acidity had lower densities and numbers of amphibian species compared to forested habitats with less acidic soils (Wyman and Jancola 1992). Because of a growing concern about the worldwide decline in amphibian species (Wake 1991) and because at least twelve amphibian species in the northeastern U.S. have exhibited population declines or range reductions in recent years (Wyman 1990), we need to be concerned about the consequences of acid deposition on amphibian conservation in the eastern deciduous forest.

Limited evidence from a study in European forests (Graveland et al. 1994) indicates that higher vertebrates, such as birds, may not be immune to the indirect effects of soil acidity caused by acid deposition. In the Netherlands, acidified, calcium-deficient soils had low populations of forest snails; these snails are an important food and calcium source of the great tit, which is a close relative of black-capped chickadees and tufted titmice of the eastern deciduous forest. Great tits that nested in forests with these acidified soils produced eggs with thin, porous shells, causing nest desertion by adults and, hence, poor reproductive performance. Similar studies are needed in the eastern deciduous forest to confirm whether acidic soil conditions created by acid deposition affect populations of forest birds. As mentioned earlier, acidic conditions can reduce aquatic insects used as food by ducklings, such as those of black and ring-necked ducks, which then can lower duckling survival (McAuley and Longcore 1988; Rusch et al. 1989).

## Other Environmental Contaminants

Excess carbon dioxide, sulfates, and other chemicals associated with the greenhouse effect and acid deposition are not the only chemicals posing threats to the eastern deciduous forest and its resources. Some chemicals intended for specific purposes, for example, to control insects in forests or on agricultural crops, have entered ecosystems and have had detrimental, long-lasting effects on various components of the ecosystem. A prime example is the insecticide DDT, which gained wide use in the 1940s but was banned in the United States in 1972 because of its impact on wildlife (Bolen and Robinson 1995). DDT entered ecological food chains throughout the world, causing major declines in numerous wildlife species. In populations of bald eagles and peregrine falcons throughout the eastern deciduous forest, DDT (and especially its metabolite DDE) accumulated in adult birds and resulted in dramatic reductions in eggshell thickness and reproductive success over several decades (Ratcliffe 1967; Sprunt 1969). Even today, despite the ban on DDT, this insecticide continues to be a concern for migratory wildlife, such as peregrine falcons, that overwinter in Central America where DDT is still used (Grier and Fyfe 1987; Steidl et al. 1991; Henny, Seegar, and Maechtle 1996). DDT has also been suspected as a probable factor causing mortality in insectivorous bats, such as the Brazilian free-tailed bat of the southern states (Clark and Kroll 1977; Rusch et al. 1989). Unlike some wildlife species, bats in the eastern deciduous forest are particularly susceptible to insecticides because they feed on large quantities of insects in order to meet high energy demands associated with flight and small size (Clark 1988).

Numerous other environmental contaminants, most notably polychlorinated biphenyls (PCBs), lead, and mercury, have found their way into terrestrial and aquatic ecosystems with serious ramifications for the health of wildlife in the eastern deciduous forest (see review by Ensor, Pitt, and Helwig 1993). Predators at the top of the food chain are particularly susceptible to these contaminants (Ensor, Pitt, and Helwig 1993). PCBs have been banned for use in the United States since 1976 in electrical transformers and electrical equipment, but considerable quantities of these chemicals remain in forest soils and in sediments of lakes in the eastern United States (Miller 1990). Lead shotgun pellets, which are ingested by ducks and geese while feeding on lake bottoms, have caused countless deaths in waterfowl and in bald eagles that feed on waterfowl crippled by lead shot (Jacobson, Carpenter, and Novilla 1977; Sanderson and Bellrose 1986). Fortunately, use of lead shot in hunting has been phased out by the U.S. Fish and Wildlife Service as of 1991. But lead exposure will continue to be a source of waterfowl mortality because it has built up in the environment over the years and because lead shot is still used in many parts of Canada (DeStefano, Brand, and Samuel 1995). However, lead, which has been called the number one environmental threat to children in the United States, still finds its way into the environment from paint used before 1978; it is also emitted into the atmosphere from leaded gasoline, fossil-fuel burning, melting and refining, waste incineration, and manufacturing processes (Miller 1992).

Although concentrations of DDT, PCBs, and lead in the environment continue to be present but are generally on the decline, mercury levels have increased in some regions of the eastern deciduous forest. Like acid deposition, the primary source of mercury is from regional atmospheric pollution, caused mainly by fossil-fuel burning, fungicides added to latex paints, and incineration of municipal wastes (Swain and Helwig 1989; Nater and Grigal 1992). Atmospheric deposition of mercury in Minnesota and Wisconsin, for example, has increased at least 3.5-fold since the 1850s (Swain et al. 1992). Mercury levels varied regionally in organic litter and forest soils but tended to increase more easterly along a continuum of upland forested sites from northwestern

Minnesota to eastern Michigan (Nater and Grigal 1992). In Minnesota, mercury levels have increased 3 to 5 percent per year in fish in recent years (Swain and Helwig 1989).

As mercury enters aquatic habitats and fish, serious problems can occur in predator populations that feed on these fish. Predation on mercury-contaminated fish by common loons in Minnesota has resulted in impaired reproduction and increased mortality (Ensor, Helwig, and Wernmer 1992). Hence, the conservation of loons may be in jeopardy because more than 50 percent of common loons in the continental United States breed in Minnesota. High levels of mercury can be deleterious or fatal to other fish eaters, such as mink and northern river otter (Wobeser and Swift 1976; Wren 1985). Mercury concentrations in Florida bald eagles have been below those causing death but are high enough to negatively affect behavior and reproduction (Wood et al. 1996). Furthermore, mercury contamination of aquatic habitats has become a human health problem because dangerous levels of mercury can occur in fish consumed by humans (Nater and Grigal 1992).

## Ozone

Ozone is vital to plant and animal life on the earth. It is produced naturally in the stratosphere by the action of solar radiation on oxygen and closer to the earth in the troposphere by electrical discharges of lightning during thunderstorms. As mentioned earlier, stratospheric ozone acts as a protective barrier against harmful UV radiation. Serious problems can potentially occur in the eastern deciduous forest when concentrations of ozone either increase above normal levels in the troposphere or become depleted in the stratosphere.

### *Effects of High Ozone Levels in the Troposphere*

Tropospheric ozone, rather than acid deposition, is considered to be the principal and most widespread air pollutant affecting forests in the United States (Chappelka 1987; Skelly et al. 1989, 1997; deSteiguer, Pye, and Love 1990; Taylor, Johnson, and Anderson 1994). Although ozone has been monitored in the United States only since the 1960s, data collected from 1985–89 indicate that year-to-year concentrations are relatively high along the East coast (U.S. Department of Agriculture 1992). This includes an area extending roughly from Georgia northward through

North and South Carolina, Virginia, eastern West Virginia, Maryland, Delaware, southeastern Pennsylvania, and southern New Jersey.

High concentrations of ozone can form near the ground when sunlight interacts with emissions of primary pollutants, such as sulfur dioxides from coal-burning power plants (Davis, Hutnik, and McClenahen 1993). Concentrations of ozone in the forest often are greater at higher than at lower elevations (Skelly, Chevone, and Yang 1982; Hildebrand 1994). High levels of ozone generally occur with air masses that develop from high pressure systems, which then transport ozone considerable distances from the source of the emissions (Wolff et al. 1977; Skelly et al. 1989). Ozone may be transported to remote sites as far as 600 to 800 kilometers from urban sources (Evans et al. 1983). If an air mass containing high concentrations of ozone is transported but becomes stationary over a forest, it can injure and retard the growth of many plant species. In a study at Great Smoky Mountains National Park, for example, more than seventy species of native plants were injured by ozone (Hacker and Renfro 1992).

The highest levels of ozone near ground level usually occur in summer, coinciding with active growth of forest vegetation (Chappelka 1987). Ozone enters a plant on the underside of leaves through small pores called stomata (Hacker and Renfro 1992). Visible symptoms of ozone injury develop within a few hours or days; leaves become discolored as cells on the upper surfaces die, thereby reducing photosynthesis in the plant (Skelly et al. 1989, 1997). The effects of ozone exposure can differ among species. Some, such as red maple and black cherry, are very sensitive to ozone injury, whereas others, such as sugar maple and white oak, show no visible effects (Davis and Skelly 1992; Hacker and Renfro 1992). In Shenandoah National Park, ozone was implicated not only as the cause of stippling on the upper leaf surfaces of black cherry, tulip poplar, and white ash but also perhaps as a factor leading to overall stress and weakening of individual trees (Hildebrand 1994) (Figure 9.10). This added stress and weakening of trees induced by ozone injury can then impair the ability of an individual tree to withstand other stress factors in the environment (Hildebrand, Skelly, and Fredericksen 1996). Ozone can affect the chemical composition of leaves, thereby making certain species, such as white oak, more "palatable" to gypsy moths and more susceptible to defoliation by this insect pest (Jeffords and Endress 1984). However, the long-term consequences of continuous exposure to tropospheric ozone on forests in the eastern United States are not yet understood (Skelly et al. 1997).

*Effects of Low Ozone Levels in the Stratosphere*

Natural events may lower levels of stratospheric ozone, as did the eruption of Mount Pinatubo in the Philippines in 1991, which expelled considerable amounts of suspended particles into the atmosphere. In particular, high concentrations of chlorine and bromine can also deplete ozone in the stratosphere (Montzka et al. 1996). Chlorine and bromine are released by UV radiation from CFCs and bromine-containing halocarbons (e.g., methyl bromide), respectively, which are chemical compounds referred to as halogens. Chlorine and bromine speed up the breakdown of ozone into oxygen gas, thereby causing a thinning of the ozone layer in the stratosphere.

In 1985, a very large loss of ozone was discovered over Antarctica. The area of loss is equivalent to the area of the continental United States (Mahlman 1992). In September 1989, ozone levels over the Southern Hemisphere were dropping at approximately 1.5 percent per day (World Resources Institute 1990). Ozone levels in the stratosphere over the Antarctic were believed to have declined 60 to 95 percent below normal in the late 1980s (Miller 1990). A comparable ozone loss has not been found over the Arctic, in part, because the atmosphere over the Northern Hemisphere is more dynamic than over the Southern

Figure 9.10. Injury to tulip poplar leaves is due to ambient ozone exposure. (Photograph by Donald D. Davis, Department of Plant Pathology, Pennsylvania State University; from Skelly et al. 1989.)

Hemisphere (Mahlman 1992). However, a hole in the ozone layer is expected to appear soon over North America because ozone levels in the stratosphere have continued to decline below normal levels in recent years. About a 3 percent decline in ozone in the stratosphere has occurred in the Northern Hemisphere since the late 1960s (Miller 1990).

With the thinning of the stratospheric layer of ozone, greater quantities of tissue-damaging UV radiation will reach the earth and potentially have far-reaching impacts on the world's plants and animals. For example, the productivity of small plant life, termed phytoplankton, which is vital to the food chain, has already been affected detrimentally by ozone depletion over the Antarctic Ocean (Weber 1994). Many baleen whales in the Antarctic Ocean rely on plankton as a food resource (Vaughan 1986). If UV radiation were to increase in the eastern deciduous forest, we could expect some negative impact on microorganisms found in forest lakes, ponds, and streams.

Vertebrates may also be susceptible to reduced levels of ozone in the stratosphere. Evidence for this is beginning to come forth. For instance, the hatching success of eggs of the western toad and the Cascades frog, two amphibian species from the mountains of Oregon, was lowered markedly by experimental doses of UV radiation (Blaustein et al. 1994). Because populations of both of these amphibian species have dropped precipitously in the wild in recent years, some scientists have speculated that excessive UV radiation caused by ozone depletion in the stratosphere over the Pacific Northwest may be partially to blame for this decline. On the other hand, UV radiation was an unlikely cause of declining populations of red-legged frogs in Oregon (Blaustein et al. 1996). Future studies of this tenuous relationship between amphibian populations and UV radiation may shed additional insight into whether any thinning of the stratospheric ozone affects other species of amphibians, especially those that occur in the eastern deciduous forest. Future research will probably show that UV radiation, habitat loss (Blaustein 1994), acid deposition (Freda, Sadinksi, and Dunson 1991), introduced predators (Fisher and Shaffer 1996; Tyler et al. 1998), and epidemic diseases (Laurance, McDonald, and Speare 1996) are among the major factors causing catastrophic population declines in amphibians throughout the world.

Another alarming by-product of ozone depletion in the stratosphere has direct implications for human health. Each 1 percent drop in ozone in the stratosphere permits 2 percent more UV radiation to reach the earth's surface, which can result in a 5 to 7 percent increase in human skin cancer (Miller 1990). A 5 percent decline in ozone in the stratosphere is expected to cause one million cases of skin cancer in the United States, plus a dramatic upsurge in incidences of eye cataracts.

The good news is that the Clean Air Act of 1990, which was passed by the U.S. Congress and enacted in 1995, will reduce airborne pollutants, such as sulfur dioxides, in the atmosphere over the eastern deciduous forest for years to come (Miller 1992). The abundance of two ozone-depleting gases, chlorine and bromine, is expected to reach a maximum in the stratosphere by 1999 and decline thereafter if limits outlined by the international agreement developed in 1987, known as The Montreal Protocol on Substances That Deplete the Ozone Layer, are not exceeded beyond 1999 (Montzka et al. 1996). International restrictions imposed by future meetings of The Montreal Protocol on Substances That Deplete the Ozone Layer will continue to help with the serious problem of worldwide depletion of stratospheric ozone (Kerr 1996). Organizations, such as Man and the Biosphere Program, which is coordinated through the United Nations Educational, Scientific, and Cultural Organization and which consists of more than 110 member countries, will help monitor and, one hopes, take the lead in mitigating global climatic change and its effects on biodiversity in forests worldwide (Laarman and Sedjo 1992).

In addition, the United States and most developed countries phased out the use and manufacture of CFCs and other halogens in the 1990s. CFCs have been replaced by hydrofluorocarbons (HFCs); these newer chemicals are perhaps fifty thousand times less destructive to ozone compared to CFCs and are expected to last only about fifteen years in the atmosphere. Unfortunately, even if all CFCs were banned immediately worldwide, ozone in the earth's stratosphere would take several decades to recover from current levels of depletion (Miller 1990). Furthermore, production of bromine-containing halocarbons was halted in 1994, but methyl bromide continues to be used as a soil and stored-grain fumigant (Kerr 1996). Although bromine is scarcer than chlorine in the stratosphere, it is forty times more damaging to ozone compared to chlorine (Kerr 1996; Montzka et al. 1996). No suitable replacements have been identified for some bromine-containing halocarbons, and the use of methyl bromine will not be phased out until at least 2010 (Kerr 1996).

In conclusion, the eastern deciduous forest and its resources will continue to be exposed to global climatic changes and airborne pollutants. Much research is needed to establish causal links between these atmospheric factors and their potential effects on forest health and biodiversity. A concerted and comprehensive understanding of the possible atmospheric influences on forest plants and animals will undoubtedly be important to ensure conservation of resources in the eastern deciduous forest for coming generations.

# 10. FORESTS OF THE FUTURE: CHALLENGES AND OPPORTUNITIES

The eastern deciduous forest is dynamic. The distribution and abundance of plant and animal species in today's forest will change over time because of myriad environmental factors, some within the control of natural resource managers (e.g., silvicultural practices) and others beyond direct control (e.g., climatic events). Certain environmental problems and issues currently affecting the eastern deciduous forest will be present well into the future; new problems and issues will surely arise. Despite changes, problems, and issues confronting the eastern deciduous forest of the twenty-first century, the major conservation goal of natural resource managers and society in general must be to ensure a healthy, productive, and sustainable forest (Durning 1994). The future management of forests will require a collegial and straightforward discussion and synthesis of forest health (DellaSalla et al. 1995).

Forest health is meant to be more than a simple assessment of the condition of trees in a forested tract. It extends to comprehensive development of a management strategy that includes a broad spectrum of health and sociopolitical issues. In other words, we must "see the forest through the trees." We can be somewhat optimistic that this overriding goal will be achieved because of an ever-expanding understanding of the ecology of the eastern deciduous forest. Moreover, such a goal will be attainable when cooperative action is promoted and achieved among natural resource managers, industry, landowners, recreationists, and environmentalists (Thorne 1993).

At least five broad and, in some cases, interrelated challenges and opportunities lay ahead for the eastern deciduous forest: biodiversity conservation and ecosystem management, forest fragmentation, education, recreation, and regional and global influences.

## Biodiversity Conservation and Ecosystem Management

Until the later part of the twentieth century, maintenance of biodiversity was not a major concern within the wildlife profession. The primary focus was on the conservation of single species, such as the white-tailed deer, American beaver, eastern wild turkey, and other important game species. We can still, however, be proud that past wildlife managers, agencies, and universities throughout the eastern deciduous forest had the willingness and expertise to save these aesthetically and economically important species from extinction.

As we move through the early decades of the twenty-first century, both single-species and multispecies approaches have a rightful place in wildlife conservation (Yahner 1990). A single-species approach is necessary to the conservation of nongame (nonconsumptive) species with specialized habitat requirements, such as the Kirtland's warbler and Appalachian woodrat. Certain game (consumptive) species, such as the ruffed grouse, a popular upland game bird, also warrant additional research and management using a single-species approach; for instance, we have only recently tested whether habitat management for ruffed grouse results in higher population densities (McDonald, Palmer, and Storm 1994).

A multispecies approach to wildlife conservation makes great sense. Considerable international focus is now being devoted to ensuring conservation of numerous Neotropical migrants. Some of these birds, for example, the scarlet tanager, require extensive forest-interior habitats, and others, such as the chestnut-sided warbler, rely on younger forests that are reverting to mature ones. Furthermore, acquisition, preservation, or management of forest land for a group of species reduce costs and labor and go hand in hand with ecosystem management efforts.

Natural resource managers from several disciplines, including forest and wildlife managers, must contribute their knowledge, experience, and understanding to make major strides toward effective ecosystem management (see DeBell and Curtis 1993). Biodiversity conservation in the eastern deciduous forest and elsewhere hinges largely on the successful implementation of ecosystem management (Grumbine 1997). Ecosystem management is the basis of a new land ethic that can help safeguard the biodiversity and ecological sustainability of the eastern deciduous forest (Wood 1994). Moreover, conservation of resources in the eastern deciduous forest must be approached from a multiresource perspective that views the forest as an interactive system of plants, animals, soil, water, and climate (Behan 1990; Ticknor 1992).

Successful biodiversity conservation and ecosystem management by the natural resource profession in the eastern deciduous forest will be best achieved when these concepts are clearly understood, put into proper perspective, and, even more important, communicated effectively to society (Samson and Knopf 1993; Grumbine 1994; Coder 1995). Achieving these goals will be a challenge because a clear and workable definition of biodiversity, let alone of biodiversity conservation or ecosystem management, has taken time to evolve within the natural resource profession (Yahner 1993b; Grumbine 1997). Another challenge is that we are in an era in which most conservation agencies are charged or concerned with conserving biodiversity but, unfortunately, are constrained by inadequate resources (monies, time, labor, and so on). Hence, ecosystem management now becomes an attractive alternative to traditional resource management, because it stresses partnerships among agencies and other stakeholders rather than agency control of management actions (Knight and Meffe 1997). Additionally, the advent of ecosystem management has given natural resource managers an opportunity to reexamine approaches and philosophies for biodiversity conservation (Czech and Krausman 1997). Conservation of biodiversity and development of a scientific basis for ecosystem management are essential to the health and productivity of the eastern deciduous forest (Society of American Foresters 1992; Scavia, Ruggiero, and Hawes 1996) and to the well-being and quality of life of all who treasure outdoor experiences in this forest.

## Forest Fragmentation

Forest fragmentation in the eastern deciduous forest will continue to be an issue. More research is required to determine its effects on sensitive species, such as Neotropical migratory birds that rely on forest interiors for nesting (Robbins 1988). Studies are needed to assess the potential impact on forest biota of land-use changes that result in forest fragmentation (e.g., the establishment of electric transmission-line corridors), preferably before such habitat changes take place (Yahner, Bramble, and Byrnes 1993). Conversely, a clearer understanding of the consequences of farm abandonment and conversion to forest on species that depend on brushy habitats or younger forests will be necessary (Litvaitis 1993).

A more complete appreciation of wildlife-landscape relationships will demand continued use of sophisticated technology, such as remote sensing and geographic information systems, to document landscape patterns and processes (Vogelmann 1995; Ross 1996; Yahner et al. 1996). Moreover, studies on wildlife-landscape relationships must be long term, with the intent of elucidating causal factors affecting biodiversity patterns and processes in the eastern deciduous forest (Yahner 1995a). Relatively recent studies conducted by conservation and wildlife biologists have made tremendous strides in explaining these relationships, many of which stemmed from concepts proposed by early island biogeography theory. Some of these studies have also urged caution when extrapolating results obtained from a landscape-wildlife study in a given region or landscape to that of another. For instance, as we have seen in chapter 7, rates of nest parasitism by cowbirds in the Midwest can dramatically differ from those in the Northeast, or rates of nest parasitism in agricultural landscapes may vary markedly from those in forested landscapes.

In a few decades, we are likely to lose most or perhaps all old-growth forests in temperate and tropical regions, largely because of forest fragmentation. We cannot allow wildlands, particularly the remaining old-growth forests, to be irresponsibly destroyed. The hope is that many forest stands will be managed on very long rotations, and others will be left unharvested to provide an emerging source of old-growth stands in all forest regions of the United States (see Ticknor 1992). Most existing stands in the eastern deciduous forest unfortunately will take from one hundred to two hundred more years of natural development to attain the plant species composition, vegetative structure (e.g., large trees and snags, fallen logs), and animal species that are characteristic of old-growth forests (Lorimer and Frelich 1994). Old-growth forests will help ensure the conservation of all biota, from forest fungi to vertebrates like the black bear, in the eastern deciduous forest. Hence, resource managers and society should be willing to set aside unexploited forested lands (Noss 1991). Resource managers must define, inventory, and map remaining old-growth stands, preserve adequate amounts of these stands to make sure that ecological interactions are maintained, and expand research to obtain more information on the ecology of these stands (The Wildlife Society 1988). State and federal agencies, which are stewards of large expanses of public lands in the eastern United States, should make a concerted effort to ensure the preservation of a certain percentage (perhaps 10 percent) of public forested lands as large tracts (> 100 hectares) for the establishment of old-growth forests.

## Education

Education will be key to conservation of the eastern deciduous forest. Old-growth forests and healthy forests in general not only are vital to biodiversity conservation but also serve as critical educational laboratories for children and students. Educating young people about the ecology and conservation of forest resources can never begin too early (Vasievich et al. 1993). When we view the eastern deciduous forest as an educational laboratory for young people, we reinforce that it is a resource now and in the future and does not merely represent a commodity to exploit (Orr 1993). The eastern deciduous forest's value as a resource will continue to increase as the eastern United States becomes more urbanized and as societal demands on the forest increase.

The Forest Stewardship Program, a landmark educational program for private forest landowners, is key to the wise use of the eastern deciduous forest. It was authorized under the 1990 Farm Bill and is administered by the U.S. Forest Service through the State and Private Forestry Program (O'Donnell 1992; Schacht 1993). The program's goal is to educate and provide assistance for private forest landowners through workshops, public service messages, and publications in order to achieve sound management of private forest lands. This Forest Stewardship Program is especially important because only a small percentage of private forest landowners hold land primarily for timber production; a greater percentage list recreation and aesthetic enjoyment as reasons for ownership (Birch 1996).

A second program established under the Farm Bill is the Stewardship Incentive Program (SIP), which complements the Forest Stewardship Program by providing incentives to implement sound forest-management goals (Figure 10.1). An attractive feature of the SIP is that it is not necessarily oriented toward timber production but can be used to protect and improve soil, water, wildlife habitat, aesthetics, and recreational opportunities. In the northeastern United States, the SIP spent more than $4.2 million in incentives to implement forest-management goals (Schacht 1993).

One potential obstacle to conservation of forest resources for educational and other purposes is continued human population growth. The human population has nearly tripled in the United States since the late nineteenth century (MacCleery 1992). Moreover, the proportion of people from a rural (farm) rather than a nonfarm background has declined. As a consequence of this population shift, students trained as

Figure 10.1. The Forest Stewardship Program and the Stewardship Incentive Program help achieve sound management of private forestlands. (Photograph by S. Williams, Agricultural Information Services, Pennsylvania State University.)

future natural resource managers and stewards of the eastern deciduous forest have less personal experience and possibly a greater spiritual separation from the forest and its resources than the typical student of a generation or two ago. This situation challenges the way elementary and secondary schools present environmental studies and the manner in which institutions of higher education approach natural resource disciplines, such as forestry and wildlife management (Meyer 1992). Yet these challenges bring new opportunities because an urbanized trend will likely enhance diversity in the natural resource profession, providing better representation by women and other minority segments of our society.

## Recreation

Since the later part of the twentieth century, recreational viewing of forest wildlife has increased dramatically in the United States (Gullion 1990; Hammitt, Dulin, and Wells 1993). A large percentage of the U.S. population participates in birding, wildlife photography, or wildlife observation (Figure 10.2). A major goal of wildlife conservation is to enhance opportunities for wildlife viewing. Providing these opportunities is important as more Americans have more time for leisure activities.

The wildlife profession is beginning to address what constitutes and determines a high-quality wildlife viewing experience. In Cades Cove, Great Smoky Mountains National Park, for instance, a wildlife viewing experience was enhanced when visitors observed many types of wildlife (including butterflies and songbirds), as well as white-tailed deer and black bear (Hammitt, Dulin, and Wells 1993). Improvement of wildlife viewing by the public coupled with educational opportunities to help viewers understand wildlife (food habits, etc.) will likely ensure a constituency that will support nongame funding and better appreciate the role of natural resource managers and agencies.

Efforts should be made to provide public access to lands for wildlife viewing experiences (e.g., see Harrington 1991). Fortunately, despite dramatic increases in the use of national parks and forests in the United States during the twentieth century, ample lands can be made available, both public and private, to meet any reasonable demands for outdoor recreation (Clawson and Harrington 1991). The negative side of providing more wildlife viewing opportunities is the potential impact on wildlife of these activities, for example, building roads (Cole and Knight 1991). Therefore, the wildlife profession has called for more research on the short- and long-term consequences of recreational use of wildlands

Figure 10.2. Recreational viewing of wildlife is a popular activity in the United States. (Photograph by S. Williams, Agricultural Information Services, Pennsylvania State University.)

on wildlife, to more effectively minimize conflicts between wildlife conservation goals and wildlands recreational use (Cole and Knight 1991; Gutzwiller 1991).

The quality of more traditional outdoor activities, such as hunting, is also in need of more study. In the northeastern United States, for example, the Delaware Water Gap National Recreation Area provides a large and diverse hunting area for residents of Pennsylvania and New Jersey (Strauss et al. 1989). Most hunters return yearly to the area and have reported crowding as the only negative feature. Hunters using this area each add $30 per activity day to the local economy, and as a group have a total yearly outlay of approximately $1.4 million in hunting expenditures.

## Regional and Global Influences

Gone are the days when resource managers could focus solely on trees and wildlife in a localized area. Just as state, federal, and international economies and trade are interdependent, so too are some environmental problems and issues facing the eastern deciduous forest and its resources. As we saw earlier, we need to be cognizant of any regional loss of biodiversity, the importance of corridors at the landscape level to wildlife, the effects of exotic species on native fauna and flora, and other issues that go beyond the boundaries of our nearby woodlot. We can no longer ignore the potential impact of regional environmental factors, such as acid deposition that originates hundreds of kilometers away, on fauna and flora in our woodland. Global climatic change and stratospheric ozone depletion, two phenomena that once seemed remotely related to natural resource conservation, now loom as serious threats to worldwide biota. Future studies should help us gain a better understanding of their effects on forest biota. However, because these environmental problems and issues are regional or global in scope, studies must address not only ecology and conservation but also economic, social, and political ramifications (e.g., see Mather 1990).

## Concluding Remarks

The eastern deciduous forest always will be with us. But how will it look in the twenty-first century? Little doubt exists that the eastern deciduous forest in the later part of the twentieth century is in better condition than it was in the 1920s (Sedjo 1991). Mistakes made in past stewardship (or lack thereof) of the eastern deciduous forest need not, and should not, be repeated. Natural resource managers and society in general should and will have a collective (and, one hopes, cooperative) say in achieving the goal of a healthy, productive, and sustainable forest. Challenges to achieving this end will always confront us, but we owe it to our children to see that this goal is met.

The eastern deciduous forest is a magnificent landscape, rich in resources ranging from biodiversity to products useful to humans. From a more personal standpoint, it has been rich in memories for me since childhood. I hope to enjoy its beauty and solitude for many years to come.

# APPENDIX A. SCIENTIFIC NAMES OF ANIMALS AND PLANTS MENTIONED IN THE TEXT

Adelgid, Balsam Woolly *Adelges piceae*
Adelgid, Hemlock Woolly *Adelges tsugae*
Alder *Alnus* spp.
Alligator *Alligator mississippiensis*
Ash *Fraxinus* spp.
Ash, Green *Fraxinus pennsylvanica*
Ash, White *Fraxinus americana*
Aspen *Populus* spp.
Aspen, Bigtooth *Populus grandidentata*
Aspen, Quaking *Populus tremuloides*
Bacterium, Lyme Disease *Borrelia burgdorferi*
Basswood, American *Tilia americana*
Basswood, White *Tilia heterophylla*
Bat (Vespertillonidae)
Bat, Brazilian Free-tailed *Tadarida brasiliensis*
Bat, Big Brown *Eptesicus fuscus*
Bat, Fruit (Pteropodidae)
Bat, Eastern Red *Lasiurus borealis*
Bat, Silver-haired *Lasiurus noctivagans*
Bear, Black *Ursus americanus*
Bear, Grizzly (Brown) *Ursus arctos*
Beauty, Spring *Claytonia virginica*
Beaver, American *Castor canadensis*
Bee, Superfamily Apoidea
Bee, Honey *Apis* spp.
Beech *Fagus* spp.
Beech, American *Fagus grandifolia*
Beech, Gray *Carpinus caroliniana*
Birch *Betula* spp.
Birch, Mountain Paper *Betula cordifolia*
Birch, Paper *Betula papyrifera*
Birch, River *Betula nigra*
Birch, Yellow *Betula lenta*
Bison *Bos bison*
Blackberry *Rubus allegheniensis*
Bluebird, Eastern *Sialia sialis*
Bobcat *Lynx rufus*
Bobolink *Dolichonyx oryzivorus*
Bobwhite, Northern *Colinus virginianus*
Bt *Bacillus thuringiensis*

Buckeye, Yellow *Aesculus octandra*
Budworm, Spruce *Choristoneura fumiferana*
Bunting, Indigo *Passerina cyanea*
Butterfly (Lepidoptera)
Cat, Domestic *Felis catus*
Cat, Saber-toothed *Smilodon fatalis*
Cankerworm, Fall *Alsophila pometaria*
Cardinal, Northern *Cardinalis cardinalis*
Catbird, Gray *Dumetella carolinensis*
Catchfly, Royal *Silene regia*
Caterpillar (Lepidoptera)
Caterpillar, Forest Tent *Malacosoma disstria*
Cedar, Atlantic White *Chamaecyparis thyoides*
Cedar, Eastern Red *Juniperus virginiana*
Cedar, Northern White *Thuja occidentalis*
Chat, Yellow-breasted *Icteria virens*
Cherry, Black *Prunus serotina*
Cherry, Pin *Prunus pensylvanica*
Chestnut, American *Castanea dentata*
Chickadee, Black-capped *Poecile atricapillus*
Chimpanzee *Pan troglodytes*
Chipmunk, Eastern *Tamias striatus*
Condor, California *Gymnogyps californicus*
Cottontail, Eastern *Sylvilagus floridanus*
Cottontail, New England *Sylvilagus transitionalis*
Cottonwood *Populus* spp.
Cottonwood, Eastern *Populus deltoides*
Cottonwood, Swamp *Populus heterophylla*
Cow, Domestic *Bos taurus*
Cowbird, Brown-headed *Molothrus ater*
Coyote *Canis latrans*
Crow, American *Corvus brachyrhynchos*
Cuckoo, Black-billed *Coccyzus erythropthalmus*
Cuckoo, Yellow-billed *Coccyzus americanus*
Cypress, Bald *Taxodium distichum*
Dace, Blacknose *Rhinichthys atratulus*
Dandelion *Taraxacum officinale*
Darter, Snail *Percina tanasi*
Deer, Sika *Cervus nippon*
Deer, White-tailed *Odocoileus virginianus*
Dog, Domestic *Canis familiaris*
Dogwood *Cornus* spp.
Dogwood, Flowering *Cornus florida*
Dove, Mourning *Zenaidura macroura*
Duck, Black *Anas rubripes*
Duck, Ring-necked *Aythya collaris*
Duck, Wood *Aix sponsa*
Eagle, Bald *Haliaeetus leucocephalus*
Elder, Box *Acer negundo*
Elk (Wapiti) *Cervus elaphus*
Elm *Ulmus* spp.
Elm, American *Ulmus americana*
Ermine *Mustela erminea*
Falcon, Peregrine *Falco peregrinus*

Fern (Polypodiaceae)
Finch, House *Carpodacus mexicanus*
Fir, Fraser *Abies fraseri*
Fir, Silver *Abies alba*
Fleabane, Common *Erigeron strigosus*
Flicker, Northern *Colaptes auratus*
Flycatcher (Tyrannidae)
Flycatcher, Acadian *Empidonax virescens*
Flycatcher, Great Crested *Myiarchus crinitus*
Fox, Arctic *Alopex lagopus*
Fox, Gray *Urocyon cinereoargenteus*
Fox, Red *Vulpes vulpes*
Fritillary, Great Spangled *Speyeria cybele*
Frog (Anura)
Frog, Cascades *Rana cascadae*
Frog, Red-legged *Rana aurora*
Fungus, Beech Bark *Nectria coccinea faginata* and *Nectria galligena*
Fungus, Chestnut Blight *Endothia parasitica*
Fungus, Dogwood Anthracnose *Discula destructiva*
Fungus, Dutch Elm Disease *Ceratocystis ulmi*
Fungus, Gypsy Moth *Entomophaga maimaiga*
Fungus, Shoestring Root Rot *Armillaria mellea*
Gnatcatcher, Blue-gray *Polioptila caerulea*
Goldenrod *Solidago* spp.
Goosefoot *Chenopodium album*
Goshawk, Northern *Accipiter gentilis*
Grackle, Common *Quiscalus quiscula*
Grape *Vitis* spp.
Grass (Gramineae)
Grosbeak, Rose-breasted *Pheucticus ludovicianus*
Grouse, Ruffed *Bonasa umbellus*
Gull, Laughing *Larus atricilla*
Hantavirus *Hantavirus* spp.
Hare, Snowshoe *Lepus americanus*
Harrier, Northern *Circus cyaneus*
Hawk (Falconiformes)
Hawk, Broad-winged *Buteo platypterus*
Hawk, Red-tailed *Buteo jamaicensis*
Heal-all *Prunella vulgaris*
Hemlock *Tsuga* spp.
Hemlock, Eastern *Tsuga canadensis*
Hen, Heath *Tympanuchus cupido cupido*
Hickory *Carya* spp.
Hickory, Bitternut *Carya cordiformis*
Hickory, Mockernut *Carya tomentosa*
Hickory, Shagbark *Carya ovata*
Horse *Equus caballus*
Hummingbird (Trochilidae)
Hummingbird, Ruby-throated *Archilochus colubris*
Jay, Blue *Cyanocitta cristata*
Junco, Dark-eyed *Junco hyemalis*
Kestrel, American *Falco sparverius*
Kinglet, Golden-crowned *Regulus satrapa*
Laurel, Mountain *Kalmia latifolia*
Leafroller, Oak *Archips semiferanus*
Lily, Wood *Lilium philadelphicum*

Lion, Mountain *Puma concolor*
Lizard (Squamata)
Locust, Black *Robinia pseudoacacia*
Locust, Honey *Gleditsia triacanthos*
Loon, Common *Gavia immer*
Lynx, Canada *Lynx canadensis*
Magnolia, Cucumber *Magnolia acuminata*
Magnolia, Southern *Magnolia grandiflora*
Mallard *Anas platyrhynchos*
Mammoth, Woolly *Mammuthus primigeius*
Manatee *Trichechus manatus*
Maple *Acer* spp.
Maple, Red *Acer rubrum*
Maple, Silver *Acer saccharinum*
Maple, Striped *Acer pensylvanicum*
Maple, Sugar *Acer saccharum*
Mastodon *Mammut americanum*
Mink *Mustela vison*
Mite, Bee Tracheal *Acarapis woodi*
Mole (Talpidae)
Monarch *Danaus plexippus*
Moose *Alces alces*
Moth, Gypsy *Lymnatria dispar*
Mouse (Rodentia)
Mouse, Deer *Peromyscus maniculatus*
Mouse, House *Mus musculus*
Mouse, Meadow Jumping *Zapus hudsonius*
Mouse, White-footed *Peromyscus leucopus*
Nuthatch (Sittidae)
Nuthatch, White-Breasted *Sitta carolinensis*
Oak *Quercus* spp.
Oak, Black *Quercus velutina*
Oak, Blackjack *Quercus marilandica*
Oak, Bur *Quercus macrocarpa*
Oak, Cherrybark *Quercus falcata pagodaefolia*
Oak, Chestnut *Quercus prinus*
Oak, Live *Quercus virginiana*
Oak, Northern Red *Quercus rubra*
Oak, Overcup *Quercus lyrata*
Oak, Pin *Quercus palustris*
Oak, Post *Quercus stellata*
Oak, Scarlet *Quercus coccinea*
Oak, Scrub *Quercus ilicifolia*
Oak, Southern Red *Quercus falcata*
Oak, Swamp Chestnut *Quercus michauxii*
Oak, Water *Quercus nigra*
Oak, White *Quercus alba*
Oak, Willow *Quercus phellos*
Opossum, Virginia *Didelphis virginiana*
Oriole, Baltimore *Icterus galbula*
Otter, Northern River *Lontra canadensis*
Ovenbird *Seiurus aurocapillus*
Owl (Strigidae)
Owl, Great Horned *Bubo virginianus*
Owl, Northern Spotted *Strix occidentalis courina*
Palm, Buccaneer *Pseudophoenix sargentii*
Panda, Giant *Ailuropoda melanoleuca*
Panther, Florida *Puma concolor*
Pecan *Carya illinoensis*
Persimmon, Common *Diospyros virginiana*

Pheasant, Ring-necked *Phasianus colchicus*
Phoebe, Eastern *Sayornis phoebe*
Pig *Sus scrofa*
Pig, Wild (Feral) *Sus scrofa*
Pigeon, Passenger *Ectopistes migratorius*
Pine *Pinus* spp.
Pine, Jack *Pinus banksiana*
Pine, Loblolly *Pinus taeda*
Pine, Longleaf *Pinus palustris*
Pine, Pitch *Pinus rigida*
Pine, Pond *Pinus serotina*
Pine, Red *Pinus resinosa*
Pine, Sand *Pinus clausa*
Pine, Shortleaf *Pinus echinata*
Pine, Slash *Pinus elliottii*
Pine, Virginia *Pinus virginiana*
Pine, White *Pinus strobus*
Pokeweed *Phytolacca americana*
Poplar, Tulip (Yellow) *Liriodendron tulipifera*
Porcupine, Common *Erethizon dorsatum*
Puccoon, Hairy *Lithospermum caroliniense*
Quail, Japanese *Coturnix coturnix*
Raccoon *Procyon lotor*
Raspberry *Rubus* spp.
Rabbit (Leporidae)
Rat, Hispid Cotton *Sigmodon hispidus*
Rat, Norway *Rattus norvegicus*
Rattlesnake, Eastern Massasauga *Sistrurus catenatus catenatus*
Rattlesnake, Timber *Crotalus horridus*
Redstart, American *Setophaga ruticilla*
Robin, American *Turdus migratorius*
Rodent (Rodentia)
Rose, Multiflora *Rosa multiflora*
Salamander (Caudata)
Salamander, Flatwoods *Ambystoma cingulatum*
Salamander, Jefferson *Ambystoma jeffersonianum*
Salamander, Redback *Plethodon cinereus*
Salamander, Spotted *Ambystoma maculatum*
Sassafras *Sassafras albidum*
Scale, Beech *Cryptococcus fagisuga*
Sculpin, Mottled *Cottus bairdi*
Sculpin, Slimy *Cottus cognatus*
Shrew (Soricidae)
Shrew, Masked *Sorex cinereus*
Shrew, Northern Short-tailed *Blarina brevicauda*
Skunk, Striped *Mephitis mephitis*
Snail (Gastropoda)
Snake (Squamata)
Spanworm, Elm *Ennomos subsignarius*
Sparrow, Bachman's *Aimophila aestivalis*
Sparrow, Cape Sable Seaside *Ammodramus maritimus*
Sparrow, Chipping *Spizella passerina*
Sparrow, Field *Spizella pusilla*
Sparrow, Grasshopper *Ammodramus savannarum*
Sparrow, House *Passer domesticus*
Sparrow, White-throated *Zonotrichia albicollis*
Spruce *Picea* spp.
Spruce, Red *Picea rubens*

Spruce, Norway *Picea abies*
Squirrel (Sciuridae)
Squirrel, Fox *Sciurus niger*
Squirrel, Gray *Sciurus carolinensis*
Squirrel, Mt. Graham Red *Tamiasciurus hudsonicus grahamensis*
Squirrel, Northern Flying *Glaucomys sabrinus*
Squirrel, Red *Tamiasciurus hudsonicus*
Squirrel, Southern Flying *Glaucomys volans*
Starling, European *Sturnus vulgaris*
Sugarberry *Celtis laevigata*
Swallowtail, Spicebush *Papilo troilus*
Sweetgum *Liquidambar styraciflua*
Sycamore, American *Platanus occidentalis*
Tanager (Traupidae)
Tanager, Scarlet *Piranga olivacea*
Teal, Blue-winged *Anas discors*
Thistle *Cirsium* spp.
Thrasher, Brown *Toxostoma rufum*
Thrip, Pear *Taeniothrip inconsequens*
Thrush, Hermit *Catharus guttatus*
Thrush, Wood *Hylocichla mustelina*
Tick, Deer *Ixodes scapularis*
Tit, Great *Parus major*
Titmouse, Tufted *Baeolophus bicolor*
Toad (Anura)
Toad, Cane *Bufo marinus*
Toad, Western *Bufo borealis*
Towhee, Eastern *Pipilo erythrophthalmus*
Trout *Salmo* spp.
Trout, Brook *Salvelinus fontinalis*
Tupelo, Black *Nyssa sylvatica*
Tupelo, Swamp *Nyssa aquatica*
Tupelo, Water *Nyssa sylvatica biflora*
Turkey, Eastern Wild *Meleagris gallopavo*
Turtle (Testudines)
Turtle, Eastern Box *Terrapene carolina*
Turtle, Yellow-bellied Slider *Pseudemys scripta*
Veery *Catharus fuscescens*
Violet *Erythronium* spp.
Vireo (Vireonidae)
Vireo, Blue-headed *Vireo solitarius*
Vireo, Red-eyed *Vireo olivaceus*
Vireo, White-eyed *Vireo griseus*
Virus, Gypsy Moth *Baculovirus* sp.
Virus, Rabies *Lyssavirus* spp.
Vole (Muridae)
Vole, Meadow *Microtus pennsylvanicus*
Vole, Rock *Microtus chrotorrhinus*
Vole, Southern Red-backed *Clethrionomys gapperi*
Vole, Woodland *Microtus pinetorum*
Vulture (Falconiformes)
Walleye *Stizostedion vitreum*
Walnut, Black *Juglans nigra*
Warbler (Parulidae)
Warbler, Black-throated Blue *Dendroica caerulscens*
Warbler, Black-throated Green *Dendroica virens*
Warbler, Blue-winged *Vermivora pinus*

Warbler, Chestnut-sided *Dendroica pensylvanica*
Warbler, Golden-winged *Vermivora chrysoptera*
Warbler, Hooded *Wilsonia citrina*
Warbler, Kentucky *Oporornis formosus*
Warbler, Kirtland's *Dendroica kirtlandii*
Warbler, Magnolia *Dendroica magnolia*
Warbler, Mourning *Oporornis philadelphia*
Warbler, Parula *Parula americana*
Warbler, Prairie *Dendroica discolor*
Warbler, Worm-eating *Helmitheros vermivorus*
Waterthrush, Louisiana *Seiurus motacilla*
Waxwing, Cedar *Bombycilla cedorum*
Weasel *Mustela* spp.
Weasel, Least *Mustela nivalis*
Weasel, Long-tailed *Mustela frenata*
Weevil, Acorn (Curculionidae)
Whale, Baleen (Mysticeti)
Whale, Right *Eubalaena glacialis*
Willow *Salix* spp.
Wolf, Dire *Canis dirus*
Wolf, Eastern Gray *Canis lupus lycaon*
Wolf, Gray (Timber) *Canis lupus*
Wolf, Red *Canis rufus*
Woodchuck *Marmota monax*
Woodcock, American *Scolopax minor*
Woodpecker (Picidae)
Woodpecker, Downy *Picoides pubescens*
Woodpecker, Ivory-billed *Campephilus principalis*
Woodpecker, Pileated *Dryocopus pileatus*
Woodpecker, Red-bellied *Melanerpes carolinus*
Woodpecker, Red-cockaded *Picoides borealis*
Woodpecker, Red-headed *Melanerpes erythrocephalus*
Wood-pewee, Eastern *Contopus virens*
Woodrat, Appalachian *Neotoma magister*
Wren, Appalachian Bewick's *Thryomanes bewickii bewickii*
Wren, Carolina *Thryothorus ludovicianus*
Wren, House *Troglodytes aedon*
Yarrow *Achillea millefolium*
Yellowthroat, Common *Geothlypis trichas*
Yew, Pacific *Taxus brevifolia*

# APPENDIX B. GLOSSARY OF TERMS USED IN THE TEXT

**Acid deposition.** The falling to the earth's surface of acidic chemicals, termed secondary air pollutants, usually with rain or snow; sometimes referred to as acid rain or acid precipitation.

**Allelopathy.** Direct or indirect harm inflicted by one plant on another by the production and the release of chemicals into the environment.

**Atmosphere.** The three thin layers of gases surrounding the earth, consisting of an inner troposphere, middle stratosphere, and outer mesosphere.

**Biodiversity.** See **biological diversity**.

**Biological diversity.** The variety of life and its processes; includes genetic, species, community and ecosystem, and landscape diversity.

**Browse.** Twigs, shoots, and leaves of trees and other woody vegetation used as food by animals.

**Carnivore.** An animal that eats meat.

**Clear-cutting method.** An even-aged system of forest management in which all trees from a given area are removed.

**Climate.** Characteristic weather patterns of a given area.

**Community.** An assemblage of species located in a particular time and place.

**Community diversity.** The variety of communities that occur over a broad geographic region or landscape.

**Competition.** An interaction between organisms of the same or different species for access to the same resource.

**Coniferous tree.** A tree that produces cones and has needle-shaped leaves, which typically are evergreen.

**Corridor.** A strip of land that differs from the adjacent habitat on each side.

**Critical habitat.** A specific area within the geographic range of a species that contains the essential features to conserve the species and that may require special management consideration or protection.

**Deciduous tree.** A tree that sheds its leaves annually.

**Dispersal.** A process that generally involves the permanent movement of an organism from its natal site to a place where it reproduces when mature.

**Ecology.** The study of interactions that determine the distribution and abundance of organisms, including plants, animals, microbes, and people, and their relationship to each other and to their surroundings.

**Ecosystem.** An assemblage of species located in a particular time and place in relation to the nonliving (abiotic) components of the environment.

**Ecosystem diversity.** The variety of ecosystems that occur over a broad geographic region or landscape.

**Ecosystem management.** The synthesis of a knowledge of ecological relationships within political, social, and value contexts to achieve the goal of long-term protection of native ecosystems and their sustainability and integrity.

**Ecotone.** A gradual transition between two distinct landscape elements, such as two plant communities.

**Edge.** An interface or junction between two distinct landscape elements, such as two plant communities or land uses.

**Edge effect.** The biological processes that characterize an edge, often expressed in terms of greater abundance and diversity of organisms.

**Environment.** The variety of habitats or conditions that affect survival and reproduction of organisms.

**Even-aged management.** A forest-management system in which trees of approximately the same size and age are maintained, usually within a relatively large area.

**Even-aged reproduction method.** An even-aged system of forest management in which most trees are removed from a stand at the initial cutting, with the residual trees intended to increase floral diversity.

**Exotic species.** A normative species that has become established in a particular place in which it did not exist before human activities or influences.

**Extirpation.** A process in which an individual, a population, or a species becomes extinct.

**Food chain.** A series of animals linked together by food and ultimately dependent on plants.

**Forb.** Nonwoody plants, excluding grasses.

**Forest.** A land area dominated by trees and other woody vegetation.

**Forest cover type.** The classification of forest vegetation based on the dominant species of trees.

**Forest fragmentation.** A process in which a large, relatively mature forested stand is converted into one or more smaller forested tracts by human land uses.

**Forest-edge species.** A species that is located at or near a forest edge or one that performs all or most of its activities at or near a forest edge.

**Forest-interior species.** A species that is located away from a forest edge or one that performs all or most activities away from a forest edge.

**Forest succession.** The regrowth of a forest stand unaided by artificial seeding, seedling planting, or other human activities, following a natural or human-induced disturbance.

**Genetic diversity.** The variation in genetic makeup of individuals of the same species within a population or group of populations in a given area.

**Genetic drift.** A genetic change in the allelic composition of a population.

**Genome resource bank.** Collection of germ plasm (gametes), embryos, DNA, blood products and tissue of a species.

**Greenhouse effect.** The insulating effect of the earth's atmosphere caused by absorption of solar radiation by trace gases.

**Habitat.** A place where an organism lives.

**Herbivore.** An animal that exclusively eats plants.

**Home range.** An area traversed by an animal in its normal day-to-day activities.

**Inbreeding.** The mating of close relatives.

**Island-biogeography theory.** A set of biological principles dealing with the distribution and abundance of organisms on islands (terrestrial and aquatic).

**Landscape.** A mosaic of different ecosystems.

**Landscape diversity.** The diversity of landscapes within a geographic region.

**Keystone species.** A species upon which some other species within a community is dependent for survival.

**Mast.** The fruit of trees and shrubs.

**Metapopulation.** A group of interconnected populations.

**Minimum dynamic area.** The smallest area of an ecosystem needed to maintain natural disturbances and to serve internally as a source of colonization by species.

**Neotropical migratory songbird.** A songbird that overwinters in Central or South America and breeds in North America.

**Nest parasitism.** Laying of one or more eggs in the nest of another species, sometimes referred to as brood parasitism.

**Old-growth forest.** An uncut, virgin forest containing large trees that may be hundreds of years old.

**Omnivore.** An animal that consumes both plant and animal material.

**Photosynthesis.** The transformation by plants of solar energy into chemical energy.

**Pollination.** A process by which plants become fertilized and reproduce sexually.

**Population.** A group of organisms of the same species located in a particular time and place.

**Predation.** The process of killing and feeding on other animals.

**Primary production.** The accumulation by plants of energy and nutrients via photosynthesis.

**Rain shadow effect.** The relatively arid or semiarid conditions caused by the flow of prevailing surface winds on the leeward side of mountains.

**Second-growth forest.** Trees and other woody vegetation that occur with plant succession after disturbance of a mature forest.

**Seed-tree method.** An even-aged system of forest management in which all trees are removed in a given area except for a few mature, seed-producing trees.

**Selective cutting.** An uneven-aged system of forest management in which single trees or a small group of trees are removed from a stand.

**Shelterwood method.** An even-aged system of forest management in which trees in a given area are removed over time by a series of cuts.

**Silviculture.** The planting, growing, and tending of a stand of trees for later harvest.

**Snag.** A dead or dying tree.

**Species-area curve.** The relationship between numbers of species and size of habitats.

**Species diversity.** The variety of organisms.

**Species richness.** The number of species.

**Sustainability.** The long-term capacity of ecosystems to produce values for society.

**Sympatric.** The range of a population or a species overlapping, at least in part, that of another population or species.

**Territoriality.** The defense of an exclusive area, termed a territory, by an individual or a group of animals against other members of the same or different species.

**Thinning.** The removal of select trees in a forest stand before actual timber harvest.

**Two-age cutting method.** An even-aged system of forest management in which most trees are removed from a stand at the initial cutting to create a two-age stand structure.

**Uneven-aged management.** A forest-management system in which trees of different size and age are maintained, usually within a relatively large area.

**Viable population.** Individuals in a local population that are needed to reproduce and maintain the population from one generation to the next.

**Weather.** The short-term changes in temperature, precipitation, wind speed, and other atmospheric conditions.

**Wildlife.** Living organisms, except domesticated animals and plants under the direct control of humans.

## REFERENCES

Aber, J. D. 1990. Forest ecology and the forest ecosystem. Pages 119–43 in R. A. Young and R. L. Giese, eds. Introduction to forest science. John Wiley & Sons, New York.

Abrams, M. D. 1992. Fire and the development of oak forests. BioScience 42:346–53.

Abrams, M. D., and C. M. Ruffner. 1995. Physiographic analysis of witness-tree distribution (1765–1798) and present forest cover through north central Pennsylvania. Can. J. For. Res. 25:659–68.

Allen, R. E., and D. R. McCullough. 1976. Deer-car accidents in southern Michigan. J. Wildl. Manage. 40:317–25.

Allen-Wardell, G., P. Bernhardt, R. Bitner, A. Burquez, S. Buchmann, J. Cane, P. A. Cox, V. Dalton, P. Feinsinger, D. Inouye, M. Ingram, C. E. Jones, K. Kennedy, P. Kevan, H. Koopowitz, R. Medellin, S. Medellin-Morales, G. P. Nabhan, B. Pavlik, V. Tepedino, P. Torchio, and S. Walker. 1998. The potential consequences of pollinator declines on the conservation of biodiversity and stability of food crop yields. Conserv. Biol. 12:8–17.

Alpert, P. 1995. Incarnating ecosystem management. Conserv. Biol. 9:952–55.

Alt, G. L., and J. M. Gruttadauria. 1984. Reuse of black bear dens in northeastern Pennsylvania. J. Wildl. Manage. 48:236–39.

Alverson, W. S., D. M. Waller, and S. L. Solheim. 1988. Forests too deer: Edge effects in northern Wisconsin. Conserv. Biol. 2:348–58.

Ambrose, J. R, and S. R. Bratton. 1990. Trends in landscape heterogeneity along the borders of Great Smoky Mountains National Park. Conserv. Biol. 4:135–43.

American Chestnut Foundation, The. 1996. The details: The breeding program. Attachment to the Report of The American Chestnut Foundation. Meadowview Research Farm, Meadowview, Va.

Andersen, D. C., and M. L. Folk. 1993. *Blarina brevicauda* and *Peromyscus leucopus* reduce overwinter survivorship of acorn weevils in an Indiana hardwood forest. J. Mammal. 74:656–64.

Andersen, D. C., and J. A. MacMahon. 1985. The effects of catastrophic ecosystem disturbance: The residual mammals at Mount St. Helens. J. Mammal. 66:581–89.

Anderson, S. H., C. S. Robbins, J. R. Partelow, and J. S. Weske. 1981. Synthesis and evaluation of avian populations and habitat data for Pennsylvania. Final Report, Eastern Energy and Land Use Team, USDI-USFWS, Laurel, Md.

Andrén, H., and P. Angelstam. 1988. Elevated predation rates as an edge effect in habitat islands: Experimental evidence. Ecology 69:544–47.

Annand, E. M., and F. R. Thompson, III. 1997. Forest bird response to regeneration practices in central hardwood forests. J. Wildl. Manage. 61:159–71.

Anthony, R. G., L. J. Niles, and J. D. Spring. 1981. Small mammal associations in forested and old-field habitats—a quantitative comparison. Ecology 62:955–63.

Aplet, G. H., R. D. Laven, and P. L. Fiedler. 1992. The relevance of conservation biology to natural resource management. Conserv. Biol. 6:298–300.

Appel, H. M., and J. C. Schultz. 1994. Oak tannins reduce effectiveness of Thuricide *(Bacillus thuringiensis)* in the gypsy moth (Lepidoptera: Lymantriidae). J. Econ. Entomol. 87:1736–42.

Arabas, K. B. 1997. Fire and vegetation dynamics in the eastern serpentine barrens. Ph.D. dissertation, Pennsylvania State Univ., University Park.

Arnason, T., R. J. Hebda, and T. Johns. 1981. Use of plants for food and medicine by native people of eastern Canada. Can. J. Botany 59:2189–325.

Arnold, D. E., P. M. Bender, A. B. Hale, and R. W. Light. 1981. Studies of infertile, acidic Pennsylvania streams and their benthic communities. Pages 15–33 in R. Singer, ed. Effects of acidic precipitation on benthos. Amer. Benthological Soc., Hamilton, N.Y.

Ash, A. N. 1997. Disappearance and return of plethodontid salamanders to clearcut plots in the southern Blue Ridge Mountains. Conserv. Biol. 11:983–89.

Ash, A. N., and R. C. Bruce. 1994. Impacts of timber harvesting on salamanders. Conserv. Biol. 8:300–01.

Askins, R. A. 1994. Open corridors in a heavily forested landscape: impact on shrubland and forest-interior birds. Wildl. Soc. Bull. 22:339–47.

Askins, R. A., J. F. Lynch, and R. Greenberg. 1990. Population declines in migratory birds in eastern North America. Pages 1–57 in D. M. Power, ed. Current ornithology. Vol. 7. Plenum Press, New York.

Baker, J. P., J. Van Sickle, C. J. Gagen, D. R. DeWalle, W. E. Sharpe, R. F. Carline, B. P. Baldigo, P. S. Murdoch, D. W. Bath, W. A. Kretser, H. A. Simonin, and P. J. Wigington, Jr. 1996. Episodic acidification of small streams in the northeastern United States: Effects on fish populations. Ecol. Applications 6:422–37.

Baker, W. L. 1994. Restoration of landscape structure altered by fire suppression. Conserv. Biol. 8:763–69.

Balcom, B. J., and R. H. Yahner. 1994. A comparison of landscape characteristics and relative abundance of great horned owls at extant and extirpated colony sites of the threatened Allegheny woodrat in Pennsylvania. Final Rept. School of Forest Resources, Pennsylvania State Univ., University Park.

———. 1996. Microhabitat and landscape characteristics associated with the threatened Allegheny woodrat. Conserv. Biol. 10:515–25.

Balick, M. J., and R. Mendelsohn. 1992. Assessing the economic value of traditional medicines from tropical rain forests. Conserv. Biol. 6:128–30.

Barber, H. L. 1984. Eastern mixed forest. Pages 345–54 in L. K. Halls, ed. White-tailed deer ecology and management. Wildl. Manage. Inst., Stackpole Books, Harrisburg, Pa.

Barbour, A. G., and D. Fish. 1993. The biological and social phenomenon of Lyme disease. Science 260:1610–16.

Barnes, B. V. 1991. Deciduous forests of North America. Pages 219–344 in D. W. Goodall, ed. Temperate ecosystems of the world. Elsevier, New York.

Barron, E. J. 1995. Climate models: How reliable are their predictions? Consequences, Autumn:17–27.

Bartlein, P. J., C. Whitlock, and S. L. Shafer. 1997. Future climate in the Yellowstone National Park region and its potential impact on vegetation. Conserv. Biol. 11:782–92.

Barton, J. H. 1992. Biodiversity at Rio. BioScience 42:773.

Batzli, G. O. 1977. Population dynamics of the white-footed mouse in floodplain and upland forests. Amer. Midland Nat. 97:18–32.

Bawa, K. S., and H. G. Wilkes. 1992. Who will speak for biodiversity? Conserv. Biol. 6:473–74.

Bawa, K. S., S. Menon, and L. R. Gorman. 1997. Cloning and conservation of biological diversity: Paradox, panacea, or pandora's box? Conserv. Biol. 11:829–30.

Bayne, E. M., and K. A. Hobson. 1997. Comparing the effects of landscape fragmentation by forestry and agriculture on predation of artificial nests. Conserv. Biol. 11:1418–29.

Bean, M. J., and D. S. Wilcove. 1997. The private-land problem. Conserv. Biol. 11:1–2.

Beebe, G. S., and P. N. Omi. 1993. Wildland burning: The perception of burning. J. Forestry 91(9):19–24.

Behan, R. W. 1990. Multi-resource forest management: A paradigmatic challenge to professional forestry. J. Forestry 88(4):12–18.

Beier, P. 1993. Determining minimum habitat areas and habitat corridors for cougars. Conserv. Biol. 7:94–108.

———. 1995. Dispersal of juvenile cougars in fragmented habitat. J. Wildl. Manage. 59:228–37.

Beier, P., and S. Loe. 1992. A checklist for evaluating impacts to wildlife movement corridors. Wildl. Soc. Bull. 20:434–40.

Beissinger, S. R. 1990. On the limits and directions of conservation biology. BioScience 40:456–57.

Bekoff, M. 1982. Coyote. Pages 447–59 in J. A. Chapman and G. A. Feldhammer, eds. Wild mammals of North America. Johns Hopkins Univ. Press, Baltimore.

Belles-Isles, J.-C., and J. Picman. 1986. House wren nest-destroying behavior. Condor 88:190–93.

Bellrose, F. C. 1976. Ducks, geese, and swans of North America. Stackpole Books, Harrisburg, Pa.

Bendel, P. R., and J. E. Gates. 1987. Home range and microhabitat partitioning of the southern flying squirrel *(Glaucomys volans)*. J. Mammal. 68:243–55.

Bendell, J. F. 1974. Effects of fire on birds and mammals. In T. T. Kozlowski and C. E. Ahlgren, eds. Fire and ecosystems. Academic Press, New York.

Bennett, A. F. 1990. Habitat corridors and the conservation of small mammals in a fragmented forest environment. Landscape Ecol. 4:109–22.

Bennett, B. C. 1992. Plants and people of the Amazonian rainforests. BioScience 42:599–607.

Bennett, J. P., R. L. Anderson, M. L. Mielke, and J. J. Ebersole. 1994. Foliar injury air pollution surveys of eastern white pine (*Pinus strobus* L.): A review. Environ. Monitoring Assessment 30:247–74.

Bielefeldt, J., and R. N. Rosenfield. 1997. Reexamination of cowbird parasitism and edge effects in Wisconsin forest. J. Wildl. Manage. 61:1222–26.

Birch, T. W. 1996. Private forest landowners of the United States, 1994. Pages 10–20 in M. J. Baughman, ed. Symposium on nonindustrial private forests: Learning from the past, prospects for the future. Univ. of Minnesota, St. Paul.

Blake, J. G. 1991. Nested subsets and the distribution of birds on isolated woodlots. Conserv. Biol. 5:58–66.

Blancher, P. J. 1991. Acidification: Implications for wildlife. Trans. N. Amer. Wildl. Nat. Resourc. Conf. 56:195–204.

Blaustein, A. R. 1994. Chicken Little or Nero's fiddle? A perspective on declining amphibian populations. Herpetologica 50:85–97.

Blaustein, A. R., P. D. Hoffman, D. G. Hokit, J. M. Kiesecker, S. C. Walls, and J. B. Hays. 1994. UV repair and resistance to solar UV-B in amphibian eggs: A link to population declines? Proc. Natl. Acad. Sci. 91:1791–95.

Blaustein, A. R., P. D. Hoffman, J. M. Kiesecker, and J. B. Hays. 1996. DNA repair activity and resistance to solar UV-B radiation in eggs of the red-legged frog. Conserv. Biol. 10:1398–1402.

Blockstein, D. E., 1990. Toward a federal plan for biological diversity. J. Forestry 88(3):14–19.

———. 1995. A strategic approach for biodiversity conservation. Wildl. Soc. Bull. 23:365–69.

Boardman, L. A. 1997. Wildlife community structure and composition in managed forested stands of central Pennsylvania. Master's thesis, Pennsylvania State Univ., University Park.

Boardman, L. A., and R. H. Yahner. 1999. Wildlife communities associated with even-aged reproduction stands in two state forests in Pennsylvania. N. J. Appl. For. 16:1–7.

Bock, C. E., and L. W. Lepthien. 1976. Changing winter distribution and abundance of the blue jay, 1962–1971. Amer. Midland Nat. 96:232–36.

Bockheim, J. G. 1990. Forest soils. Pages 86–97 in R. A. Young and R. L. Giese, eds. Introduction to forest science. John Wiley & Sons, New York.

Böhning-Gaese, K., M. L. Taper, and J. H. Brown. 1993. Are declines in North American insectivorous songbirds due to causes on the breeding range? Conserv. Biol. 7:76–86.

Bolen, W. L., and E. G. Robinson. 1995. Wildlife ecology and management. 3rd ed. Prentice Hall, Inc., Englewood Cliffs, New Jersey.

Bollinger, E. K., and E. T. Linder. 1994. Reproductive success of Neotropical migrants in a fragmented Illinois forest. Wilson Bull. 106:46–54.

Bonan, G. B., D. Pollard, and S. L. Thompson. 1992. Effects of boreal forest vegetation on global climate. Nature 359:716–18.

Bond, R. R. 1957. Ecological distribution of breeding birds in the upland forests of southern Wisconsin. Ecol. Monogr. 27:351–84.

Bonnicksen, T. M. 1990. The development of forest policy in the United States. Pages 5–32 in R. A. Young and R. L. Giese, eds. Introduction to forest science. John Wiley & Sons, New York.

Booth, D. C. 1991. Seriousness of Lyme disease prompts effort to reduce the abundance of deer ticks. J. Forestry 89(1):27–29.

Botkin, D. P., D. A. Woodby, and R. A. Nisbet. 1991. Kirtland's warbler habitat: A possible early indicator of climate warming. Biol. Conserv. 56:63–78.

Boutin, S. 1992. Predation and moose population dynamics: A critique. J. Wildl. Manage. 56:116–27.

Bowles, G. H., and J. M. Campbell. 1993. Relationships between population density of white-tailed deer and the density of understory trees in forests of Erie County, Pennsylvania. J. Pennsylvania Acad. Sci. 67:109–14.

Boza, M. A. 1993. Conservation in action: Past, present, and future of the National Park System of Costa Rica. Conserv. Biol. 7:239–47.

Bramble, W. C., R. H. Yahner, and W. R. Byrnes. 1992. Breeding-bird population changes following rights-of-way maintenance treatments. J. Arboriculture 18:23–32.

Bratton, S. P. 1980. Impacts of white-tailed deer on the vegetation of Cades Cove, Great Smoky Mountains National Park. Proc. Ann. Conf. Southeast Assoc. Fish Wildl. Agencies 33:305–12.

Braun, E. L. 1950. Deciduous forests of eastern North America. Macmillan, New York.

Brauning, D. 1983. Nest site selection of the American kestrel *(Falco sparverius).* Raptor Res. 17:122.

———, ed. 1992. Atlas of breeding birds in Pennsylvania. Univ. of Pittsburgh Press, Pittsburgh.

Brawn, D., W. H. Elder, and K. E. Evan. 1982. Winter foraging by cavity nesting birds in an oak-hickory forest. Wildl. Soc. Bull. 10:271–75.

Brazier, J. R., and G. W. Brown. 1973. Buffer strips for stream temperature control. Res. Paper 15, School of Forestry, Oregon State Univ., Corvallis.

Brender, E. V., and R. W. Cooper. 1968. Prescribed burning in Georgia's Piedmont loblolly pine stands. J. Forestry 66:31–36.

Brennan, L. A. 1991. How can we reverse the northern bobwhite population decline? Wildl. Soc. Bull. 19:544–55.

Briggs, J. M., and K. G. Smith. 1989. Influence of habitat on acorn selection by *Peromyscus leucopus.* J. Mammal. 70:35–43.

Brittingham, M. C., and S. A. Temple. 1983. Have cowbirds caused forest songbirds to decline? BioScience 33:31–35.

Brodsky, L. M., and R. J. Weatherhead. 1984. Behavioral and ecological factors contributing to American black duck-mallard hybridization. J. Wildl. Manage. 48:846–52.

Brooks, M. G. 1951. Effects of black walnut trees and their products on other vegetation. West Virginia Agric. Exp. Stat. Bull. 347, Morgantown.

Brose, R. H., and L. H. McCormick. 1992. Effects of prescribed fire on pear thrips in Pennsylvania sugarbushes. N. J. Appl. For. 9:157–60.

Brothers, T. S., and A. Spingarn. 1992. Forest fragmentation and alien plant invasion of central Indiana old-growth forests. Conserv. Biol. 6:91–100.

Brown, G. W., and J. T. Krygier. 1967. Changing water temperature in a small mountain stream. J. Soil Water Conserv. 22:242–44.

Brussard, P. F., D. D. Murphy, and R. F. Noss. 1992. Strategy and tactics of conserving biological diversity in the United States. Conserv. Biol. 6:157–59.

Bubier, J. L., T. R. Moore, and N. T. Roulet. 1993. Methane emissions from wetlands in the midboreal region of northern Ontario, Canada. Ecology 74:2240–54.

Burdick, D. M., D. Cushman, R. Hamilton, and J. G. Gosselink. 1989. Faunal changes and bottomland hardwood forest loss in the Tensas Watershed, Louisiana. Conserv. Biol. 3:282–92.

Burger, G. V. 1973. Practical wildlife management. Winchester Press, New York.

Campbell, J. M. 1993. Effects of grazing by white-tailed deer on a population of *Lithospermum caroliniense* at Presque Isle. J. Pennsylvania Acad. Sci. 67:103–08.

Campbell, L. A., J. G. Hallett, and M. A. O'Connell. 1996. Conservation of bats in managed forests: Use of roosts by *Lasionycteris noctivagans*. J. Mammal. 77:976–84.

Campbell, R. W., and R. J. Sloan. 1977. Forest stand responses to defoliation by gypsy moths. For. Sci. Monogr. 19:1–34.

Carey, A. B. 1982. The ecology of red foxes, gray foxes, and rabies in the eastern United States. Wildl. Soc. Bull. 10:18–26.

Carey, A. B., and R. O. Curtis. 1996. Conservation of biodiversity: A useful paradigm for forest ecosystem management. Wildl. Soc. Bull. 24:610–20.

Carroll, C. R. 1992. Ecological management of sensitive natural areas. Pages 347–72 in P. L. Fiedler and S. Kjain, eds. Conservation biology: The theory and practice of nature conservation, preservation, and management. Chapman & Hall, New York.

Carroll, R., C. Augspurger, A. Dobson, J. Franklin, G. Orians, W. Reid, R. Tracy, D. Wilcove, and J. Wilson. 1996. Strengthening the use of science in achieving the goals of the Endangered Species Act: An assessment by the Ecological Society of America. Ecol. Applications 6:1–11.

Casey, D., and D. Hein. 1983. Effects of heavy browsing on a bird community in deciduous forest. J. Wildl. Manage. 47:829–36.

Caughley, G., and A. R. E. Sinclair. 1994. Wildlife ecology and management. Blackwell Scientific Publ., Boston.

Centers for Disease Control and Prevention. 1991. Rabies prevention—United States, 1991, recommendations of the Immunizations Practices Advisory Committee (ACIP). Morbidity Mortality Weekly Rept. 40(RR-3):1–9.

Chapman, J. A., J. G. Hockman, and W. R. Edwards. 1982. Pages 83–123 in J. A. Chapman and G. A. Feldhammer, eds. Wild mammals of North America. Johns Hopkins Univ. Press, Baltimore.

Chapman, R. C. 1978. Rabies: Decimation of a wolf pack in arctic Alaska. Science 201:365–67.

Chappelka, A. H. 1987. Air pollution: A threat to forests in the South? Alabama Forests 31(6):31–32.

Chasko, G. G., and J. E. Gates. 1982. Avian habitat suitability along a transmission-line corridor in an oak-hickory forest region. Wildl. Monogr. 82:1–41.

Childs, J. E., J. N. Mills, and G. E. Glass. 1995. Rodent-borne hemorrhagic fever viruses: A special risk for mammalogists? J. Mammal. 76:664–80.

Christensen, N. L., A. M. Bartuska, J. H. Brown, S. Carpenter, C. D'Antonio, R. Francis, J. F. Franklin, J. A. MacMahon, R. F. Noss, D. J. Parsons, C. H. Peterson, M. G. Turner, and R. G. Woodmansee. 1996. The report of the Ecological Society of America Committee on the Scientific Basis for Ecosystem Management. Ecol. Applications 6:665–91.

Cinq-Mars, J., A. W. Diamond, and E. T. LaRoe. 1991. The effect of global climate change on fish and wildlife resources. Trans. N. Amer. Wildl. Res. Conf. 56:171–76.

Clark, D. R., Jr. 1988. How sensitive are bats to insecticides? Wildl. Soc. Bull. 16:399–403.

Clark, D. R., Jr., and J. C. Kroll. 1977. Effects of DDE on experimentally poisoned free-tailed bats *(Tadarida brasiliensis)*: Lethal brain concentrations. J. Toxicol. Environ. Health 3:893–901.

Clark, J. 1993. The National Wildlife Refuge System's role in conservation of Neotropical migratory birds. Trans. N. Amer. Wildl. Nat. Res. Conf. 58:408–16.

Clawson, M. 1979. Forests in the long sweep of U.S. history. Science 204:1168–74.

Clawson, M., and W. Harrington. 1991. The growing role of outdoor recreation. Pages 249–82 in K. D. Frederick and R. A. Sedjo, eds. America's renewable resources: Historical trends and current challenges. Resources for the Future, Washington, D.C.

Coder, K. 1995. Defining ecosystem management: Embracing and delineating an educational role. Pages 85–94 in B. Hubbard, ed. Education and communication applications in natural resource management. Univ. of Georgia, Athens.

Cohn, J. R 1989. Gauging the biological impacts of the greenhouse effect. BioScience 39:142–46.

———. 1993. The national biological survey. BioScience 43:521–22.

Coile, T. S. 1937. Distribution of forest tree roots in North Carolina soils. J. Forestry 35:247–57.

Coker, D. R., and D. E. Capen. 1995. Landscape-level habitat use by brown-headed cowbirds in Vermont. J. Wildl. Manage. 59:631–37.

Cole, D. N., and R. L. Knight. 1991. Wildlife preservation and recreational use: Conflicting goals of wildland management. Trans. N. Amer. Wildl. Nat. Resourc. Conf. 56:233–37.

Collier, B. D., G. W. Cox, A. W. Johnson, and P. C. Miller. 1973. Dynamic ecology. Prentice-Hall, Englewood Cliffs, N.J.

Conner, R. N., J. W. Via, and I. D. Pather. 1979. Effects of pine-oak clear-cutting on winter and breeding birds in southwestern Virginia. Wilson Bull. 91:301–16.

Conover, M. R. 1997. Monetary and intangible valuation of deer in the United States. Wildl. Soc. Bull. 25:298–305.

Conover, M. R., W. C. Pitt, K. K. Kessler, T. J. DuBow, and W. A. Sanborn. 1995. Review of human injuries, illnesses, and economic losses caused by wildlife in the United States. Wildl. Soc. Bull. 23:407–414.

Constantine, D. G. 1967. Rabies transmission by air in bat caves. U.S. Govt. Printing Off., Washington, D.C., Public Health Serv. Publ. 1617.

Cooter, E. J., B. K. Eder, S. K. LeDuc, and L. Truppi. 1993. Climate change: Models and forest research. J. Forestry 91(9):38–43.

Coulter, M. W., and J. C. Baird. 1982. Changing forest land uses and opportunities for management in New England and the Maritime Provinces. Pages 75–85 in T. J. Dryer and G. L. Storm, tech. coord. Woodcock ecology and management. USDA–Fish Wildl. Serv., Wild. Res. Rept. 14, Washington, D.C.

Council on Environmental Quality. 1989. Environmental trends. U.S. Government Printing Office, Washington, D.C.

Cox, P. A., T. Elmquist, E. D. Pierson, and W. B. Rainey. 1991. Flying foxes as strong indicators in South Pacific Island ecosystems: A conservation hypothesis. Conserv. Biol. 5:448–54.

Cox, R. M., G. Lemieux, and M. Lodin. 1996. The assessment and condition of Fundy white birches in relation to ambient exposure to acid marine fogs. Can. J. For. Res. 26:682–88.

Craven, S. R. 1989. In defense of a territory. Pages 168–74 in S. Atwater and J. Schell, eds. Ruffed grouse. Stackpole Books, Harrisburg, Pa.

Crawford, H. S., R. G. Hooper, and R. W. Titterington. 1981. Songbird population response to silvicultural practices in central Appalachian hardwoods. J. Wildl. Manage. 45:680–92.

Crawford, H. S., and R. W. Titterington. 1979. Effects of silvicultural practices on bird communities in upland spruce-fir stands. Pages 110–119 in R. M. DeGraaf and K. E. Evans, comps. Proc. of the workshop for the management of north-central and northeastern forests for nongame birds. USDA, For. Serv. Gen. Tech. Rept. NC-51, St. Paul, Minn.

Cronon, W. 1983. Changes in the land: Indians, colonists, and the ecology of New England. Hill and Wang, New York.

Crow, T. R. 1988. Reproductive mode and mechanisms for self-replacement of northern red oak *(Quercus rubra)*—a review. For. Sci. 34:19–40.

Cummins, C. P. 1986. Effects of aluminum and low pH on growth and development in *Rana temporaria* tadpoles. Oecologia 69:248–52.

Curtis, J. T. 1959. The vegetation of Wisconsin. Univ. of Wisconsin Press, Madison.

Curtis, P. D., and M. E. Richmond. 1992. Future challenges of suburban white-tailed deer management. Trans. N. Amer. Wildl. Nat. Resourc. Conf. 57:104–14.

Cutter, S. L., H. L. Renwick, and W. H. Renwick. 1991. Exploitation, conservation, preservation: A geographic perspective of natural resource use. 2d ed. John Wiley & Sons, New York.

Cypher, B. L., R. H. Yahner, and E. A. Cypher. 1988. Seasonal food use by white-tailed deer at Valley Forge National Historical Park, Pennsylvania, USA. Environ. Manage. 12:237–42.

Cypher, E. A., R. H. Yahner, G. L. Storm, and B. L. Cypher. 1986. Flora and fauna survey in a proposed recreational area of Valley Forge National Historical Park. Proc. Pennsylvania Acad. Sci. 60:47–50.

Czech, B., and P. R. Krausman. 1997. Implications of an ecosystem management literature review. Wildl. Soc. Bull. 25:667–75.

Daily, G. C., and P. R. Ehrlich. 1992. Population, sustainability, and Earth's carrying capacity. BioScience 42:761–71.

Dale, F. H. 1954. The influence of calcium on the distribution of the pheasant in North America. Trans. N. Amer. Wildl. Conf. 19:316–23.

———. 1955. The role of calcium in reproduction of the ring-necked pheasant. J. Wildl. Manage. 19:325–31.

Darley-Hill, S., and W. C. Johnson. 1981. Acorn dispersal by the blue jay *(Cyanocitta cristata).* Oecologia 50:231–32.

Davis, D. D., R. J. Hutnik, and J. R. McClenahen. 1993. Evaluation of vegetation near coal-burning power plants in southwestern Pennsylvania. II. Ozone injury on foliage of hybrid poplar. J. Air Waste Manage. Assoc. 43:760–64.

Davis, D. D., and J. M. Skelly. 1992. Foliar sensitivity of eight eastern hardwood tree species to ozone. Water Air Soil Pollut. 62:269–77.

Day, G. M. 1953. The Indian as an ecological factor in the northeastern forest. Ecology 34:329–46.

DeBell, D. S., and R. O. Curtis. 1993. Silviculture and new forestry in the Pacific Northwest. J. Forestry 91(12):25–30.

deCalesta, D. S. 1994. Effect of white-tailed deer on songbirds within managed forests in Pennsylvania. J. Wildl. Manage. 58:711–18.

Decker, D. J. 1987. Management of suburban deer: An emerging emergency. Proc. East. Wildl. Damage Control Conf. 3:334–45.

DeGraaf, R. M., R. A. Askins, and W. M. Healy. 1993. The use of nature's constancy—preservation, protection, and ecosystem management. Trans. N. Amer. Wildl. Nat. Res. Conf. 58:17–28.

Delcourt, H. R., and P. A. Delcourt. 1997. Pre-Columbian Native American use of fire on southern Appalachian landscapes. Conserv. Biol. 11:1010–14.

Delcourt, P. A., and H. R. Delcourt. 1981. Vegetation maps for eastern North America: 49,000 B.P. to the present. Pages 123–66 in R. Romans, ed. Proc. 1980 Geobotany Conf., New York.

DellaSalla, D. A., D. M. Olson, S. E. Barth, S. L. Crane, and S. A. Primm. 1995. Forest health: Moving beyond rhetoric to restore healthy landscapes in the inland Northwest. Wildl. Soc. Bull. 23:346–56.

DeLong, C. A., and R. H. Yahner. 1996. Predation on planted acorns in managed forested stands of central Pennsylvania. Northeast Wildl. 53:11–18.

DeLong, D. C., Jr. 1996. Defining biodiversity. Wildl. Soc. Bull. 24:738–49.

deMaynadier, P. G., and M. L. Hunter, Jr. 1998. Effects of silvicultural edges on the distribution and abundance of amphibians in Maine. Conserv. Biol. 12:340–52.

Demers, M. N., J. W. Simpson, R. E. J. Boerner, A. Silva, L. Berns, and F. Artigas.

1995. Fencerows, edges, and implications of connectivity illustrated by two contiguous Ohio landscapes. Conserv. Biol. 9:1159–68.

Derge, K. L. 1997. Habitat use by eastern fox squirrels *(Sciurus niger vulpinus)* and gray squirrels *(Sciurus carolinensis)* at forest-farmland interfaces of the Valley and Ridge Province, Pennsylvania. Master's thesis, Pennsylvania State Univ., University Park.

Derge, K. L., and R. H. Yahner. 2000. Abundance of and habitat use by sympatric fox squirrels *(Sciurus niger)* and gray squirrels *(Sciurus carolinensis)* at forest-farmland interfaces of Pennsylvania. Amer. Midland Nat. 135: (in press).

Dessecker, D. R., and R. H. Yahner. 1987. Breeding bird communities associated with Pennsylvania hardwood stands. Proc. Pennsylvania Acad. Sci. 61:170–73.

DeStefaano, S., C. J. Brand, and M. D. Samuel. 1995. Seasonal ingestion of toxic and nontoxic shot by Canada geese. Wildl. Soc. Bull. 23:502–06.

deSteiguer, J. E., J. M. Pye, and C. S. Love. 1990. Air pollution damage to U.S. forests. J. Forestry 88(8):17–22.

DeWalle, D. R. 1995. Forest hydrology. Pages 109–17 in Encyclopedia of environmental biology, Vol. 2. Academic Press, New York.

Diamond, A. W. 1991. Assessment of the risks from tropical deforestation to Canadian songbirds. Trans. N. Amer. Wildl. Nat. Res. Conf. 56:177–94.

Diefenbach, D. R., W. L. Palmer, and W. K. Shope. 1997. Attitudes of Pennsylvania sportsmen towards managing white-tailed deer to protect the ecological integrity of forests. Wildl. Soc. Bull. 25:244–51.

Dobson, A. 1992. Withering heats. Nat. Hist. 9/92:3–8.

Donnelly, J. R., J. B. Shane, and H. W. Yawney. 1991. Harvesting causes only minor changes in physical properties of an upland Vermont soil. J. Forestry 89(7):28–31.

Donovan, T. M., F. R. Thompson, III, J. Faaborg, and J. R. Probst. 1995. Reproductive success of migratory birds in habitat sources and sinks. Conserv. Biol. 9:1380–95.

Drohan, J. R., and W. E. Sharpe. 1997. Long-term changes in forest soil acidity in Pennsylvania, USA Water, Air, and Soil Pollution 95:299–311.

Duchin, D. S., F. T. Koster, C. J. Peters, G. L. Simpson, B. Tempest, S. R. Zaki, T. G. Ksiazek, P. E. Rollin, S. Nichol, E. T. Umland, R. L. Mollenaar, S. E. Reef, K. B. Nolte, M. M. Gallaher, J. C. Butler, R. F. Breiman, and The Hantavirus Study Group. 1994. Hantavirus pulmonary syndrome: A clinical description of seventeen patients with a newly recognized disease. New England J. Medicine 330:949–55.

Duffield, J. W. 1990. Forest regions of North America and the world. Pages 33–65 in R. A. Young and R. L. Giese, eds. Introduction to forest science. John Wiley & Sons, New York.

Duffy, D. C., and A. J. Meier. 1992. Do Appalachian herbaceous understories ever recover from clear-cutting? Conserv. Biol. 6:196–201.

Duguay, J. P. 1997. Influence of two-age and clear-cut timber management practices on songbird abundance, nest success, and invertebrate biomass in West Virginia. Ph.D. dissertation, West Virginia Univ., Morgantown.

Dunning, J. B., Jr., R. Borgella, Jr., K. Clements, and G. K. Meffe. 1995. Patch isolation, corridor effects, and colonization by a resident sparrow in a managed pine woodland. Conserv. Biol. 9:542–50.

Dupuis, L. A., J. N. M. Smith, and F. Bunnell. 1995. Relation of terrestrial-breeding amphibian abundance to tree-stand age. Conserv. Biol. 9:645–53.

Durning, A. T. 1994. Redesigning the forest economy. Pages 22–40 in L. Starke, ed. State of the world—1994. W. W. Norton and Co., New York.

Dyck, A. P., and R. A. MacArthur. 1993. Seasonal variation in the microclimate and gas composition of beaver lodges in a boreal environment. J. Mammal. 74:180–88.

Easterling, W. E. 1990. Climate trends and prospects. Pages 32–55 in R. N. Sampson and D. Hair, eds. Natural resources for the 21st century. Amer. For. Assoc., Island Press, Washington, D.C.

Eastman, D. S., C. Bryden, M. Eng, R. Kowall, H. Armleder, E. Lofroth, and S. Stevenson. 1991. Silviculturists and wildlife habitat managers: Competitors or cooperators? Trans. N. Amer. Wildl. Nat. Res. Conf. 56:640–51.

Eckess, E. 1982. History of spread and where is it going? Pages 11–13 in S. R. Cochran, J. C. Finley, and M. J. Baughman, eds. Proc. of coping with the gypsy moth. Penn State Forestry Issues Conf., Pennsylvania State Univ., University Park.

Edwards, J. W., and D. C. Guynn, Jr. 1995. Nest characteristics of sympatric populations of fox and gray squirrels. J. Wildl. Manage. 59:103–10.

Ehrlich, P. R., and E. O. Wilson. 1991. Biodiversity studies: Science and policy. Science 253:758–62.

Eisner, T., and E. A. Beiring. 1994. Biotic exploration fund—protecting biodiversity through chemical prospecting. BioScience 44:95–98.

Elfring, C. 1989. Yellowstone: Fire storm over fire management. BioScience 39:667–72.

Elkinton, J. S., W. M. Healy, J. P. Buonaccorsi, G. H. Boettner, A. M. Hazzard, H. R. Smith, and A. M. Liebhold. 1996. Interactions among gypsy moths, white-footed mice, and acorns. Ecology 77:2332–42.

Eller, J. H., W. G. Wathen, and M. R. Pelton. 1989. Reproduction in black bears in southern Appalachian Mountains. J. Wildl. Manage. 53:353–60.

Ellstrand, N. C., and D. R. Elam. 1993. Population genetic consequences of small population size: Implications for plant conservation. Ann. Rev. Ecol. Syst. 24:217–42.

Ensor, K. L., D. D. Helwig, and L. C. Wemmer. 1992. Mercury and lead in

Minnesota common loons *(Gavia immer).* Minnesota Pollution Control Agency, St. Paul.

Ensor, K. L., W. C. Pitt, and D. D. Helwig. 1993. Contaminants in Minnesota wildlife 1989–91. Minnesota Pollution Control Agency, St. Paul.

Erb, C. 1987. The jury is still out. Penn State Agric. Fall:2–11.

Erlinge, S., G. Göransson, G. Högstedt, G. Jansson, O. Liberg, J. Loman, L. N. Nilsson, T. von Schantz, and M. Sylvén. 1984. Can vertebrate predators regulate their prey? Amer. Nat. 123:125–33.

Esler, D., and J. B. Grand. 1993. Factors influencing depredation of artificial duck nests. J. Wildl. Manage. 57:244–48.

Evans, G. P., P. Finkelstein, B. Martin, N. Possiel, and M. Grass. 1983. Ozone measurements from a network of remote sites. J. Air Pollut. Control Assoc. 33:291–96.

Fajer, E. D. 1989. How enriched carbon dioxide environments may alter biotic systems even in the absence of climatic changes. Conserv. Biol. 3:318–30.

Feldhammer, G. A. 1982. Sika deer. Pages 1114–23 in J. A. Chapman and G. A. Feldhammer, eds. Wild mammals of North America. Johns Hopkins Univ. Press, Baltimore.

Finch, D. M. 1991. Population ecology, habitat requirements, and conservation of Neotropical migratory birds. USDA–For. Serv., Fort Collins, Colo.

Finch, D. M., W. M. Block, R. A. Fletcher, and L. E. Fager. 1993. Integrating Neotropical migratory birds into Forest Service plans for ecosystem management. Trans. N. Amer. Wildl. Nat. Res. Conf. 58:417–22.

Finley, J., S. Camazine, and M. Frazier. 1996. The epidemic of honeybee colony losses during the 1995–96 season. Amer. Bee J. 136:805–08.

Fish, D., and T. J. Daniels. 1990. The role of medium-sized mammals as reservoirs of *Borrelia burgdorferi* in southern New York. J. Wildl. Diseases 26:339–45.

Fishbein, D. B. 1991. Rabies in humans. Pages 519–49 in G. M. Baer, ed. The natural history of rabies. CRC Press, Boca Raton, Fla.

Fisher, R. N., and H. B. Shaffer. 1996. The decline of amphibians in California's Great Central Valley. Conserv. Biol. 10:1387–97.

Flavin, C., and N. Lenssen. 1994. Reshaping the power industry. Pages 61–80 in L. Starke, ed. State of the world 1994. W. W. Norton and Co., New York.

Flyger, V., D. L. Leedy, and T. M. Franklin. 1983. Wildlife damage control in eastern cities and suburbs. Proc. East. Wildl. Damage Control Conf. 1:27–32.

Forbes, S. H., and D. K. Boyd. 1996. Genetic variation of naturally colonizing wolves in the central Rocky Mountains. Conserv. Biol. 10:1082–90.

Forbes, S. E., L. M. Lang, S. A. Liscinsky, and H. A. Roberts. 1971. The white-tailed deer in Pennsylvania. Res. Bull. No. 170, Pennsylvania Game Commission, Harrisburg.

Fore, T. H. 1996. Winter colony loss reported by state apiary inspectors surveyed by American Beekeeping Federation. The Speedy Bee 25:16.
Forman, R. T. T., and M. Godron. 1986. Landscape ecology. John Wiley & Sons, New York.
Foster, M. L., and S. R. Humphrey. 1995. Use of highway underpasses by Florida panthers and other wildlife. Wildl. Soc. Bull. 23:95–100.
Foster, J. R., J. L. Roseberry, and A. Woolf. 1997. Factors influencing efficiency of white-tailed deer harvest in Illinois. J. Wildl. Manage. 61:1091–97.
Franklin, J. F. 1990. Creating alternative approaches to management. Forest Prescription, Spring:5.
———. 1993. Preserving biodiversity: Species, ecosystems, or landscapes? Ecol. Applications 3:202–05.
———. 1994. Preserving biodiversity: Species in landscapes. Ecol. Applications 4:205–07.
———. 1997. Ecosystem management: An overview. Pages 21–53 in M. A. Boyce and A. Haney, eds. Ecosystem management. Yale University, New Haven, Conn.
Franklin, J. F., D. R. Berg, D. A. Thornburgh, and J. C. Tappeiner. 1997. Alternative silvicultural approaches to timber harvesting: Variable retention harvest systems. Pages 111–39 in K. A. Kohm and J. F. Franklin, eds. Creating a forestry for the 21st century: The science of ecosystem management. Island Press, Washington, D.C.
Franklin, J. F., F. J. Swanson, M. E. Harmon, D. A. Perry, T. A. Spies, V. H. Dale, A. McKee, W. K. Ferrell, J. E. Means, S. V. Gregory, J. D. Lattin, T. D. Schowalter, and D. Larsen. 1991. Effects of global climatic change on forests in northwestern North America. Northwest Environmental J. 7:233–54.
Fraser, C. M., S. Casjens, W. M. Huang, G. G. Sutton, R. Clayton, R. Lathigra, O. White, K. A. Ketchum, R. Dodson, E. K. Hickey, M. Gwinn, B. Dougherty, J.-F. Tomb, R. D. Fleischmann, D. Richardson, J. Peterson, A. R. Kerlavage, J. Quackenbush, S. Salzberg, M. Hanson, R. van Vugt, N. Palmer, M. D. Adams, J. Gocayne, J. Weidman, T. Utterback, L. Watthey, L. McDonald, P. Artiach, C. Bowman, S. Garland, C. Fujii, M. D. Cotton, K. Horst, K. Roberts, B. Hatch, H. O. Smith, and J. C. Venter. 1997. Genomic sequence of a Lyme disease spirochaete, *Borrelia burgdorferi.* Nature 390:580–86.
Frazier, M. T., J. Finley, C. H. Collison, and E. Rajotte. 1994. The incidence and impact of honeybee tracheal mites and nosema disease on colony mortality in Pennsylvania. BeeSci. 3(2):94–100.
Freda, J. 1986. The influence of acidic pond water on amphibians: A review. Water Air Soil Pollut. 30:439–50.
Freda, J., W. J. Sadinski, and W. A. Dunson. 1991. Long-term monitoring of

amphibian populations with respect to the effects of acid deposition. Water Air Soil Pollut. 55:445–62.

Friesen, L. E., P. F. J. Eagles, and R. J. MacKay. 1995. Effects of residential development on forest-dwelling Neotropical migrant songbirds. Conserv. Biol. 9:1408–14.

French, C. E., L. C. McEwen, N. D. Magruder, R. H. Ingram, and R. W. Swift. 1956. Nutrient requirements for growth and antler development in the white-tailed deer. J. Wildl. Manage. 20:221–32.

Frenzen, P. 1992. Mount St. Helens: A laboratory for research and education. J. Forestry 90(5):12–18, 37.

Frey, J. K. 1992. Response of a mammalian faunal element to climatic change. J. Mammal. 73:43–50.

Frissell, C. A., R. K. Nawa, and R. Noss. 1992. Is there any conservation biology in "New Perspectives"?: A response to Salwasser. Conserv. Biol. 6:461–64.

Fujita, M. S., and M. D. Tuttle. 1991. Flying foxes (Chiroptera: Pteropodidae): Threatened animals of key ecological and economic importance. Conserv. Biol. 5:455–63.

Gage, K. L., R. S. Ostfeld, and J. G. Olson. 1995. Nonviral vector-borne zoonoses associated with mammals in the United States. J. Mammal. 76:695–715.

Gale, R. P, and S. M. Cordray. 1991. What should forests sustain? Eight answers. J. Forestry 89(5):31–36.

Galli, A. E., C. L. Leck, and R. T. T. Forman. 1976. Avian distribution patterns in forest islands of different sizes in central New Jersey. Auk 93:356–64.

Galligan, J. H., and W. A. Dunson. 1979. Biology and status of timber rattlesnake *(Crotalus horridus)* populations in Pennsylvania. Biol. Conserv. 15:13–58.

Galloway, J. N., G. E. Likens, W. C. Keene, and J. M. Miller. 1982. The composition of precipitation in remote areas of the world. J. Geophys. Res. 87:8771–86.

Gardner, A. L. 1982. Virginia opossum. Pages 3–36 in J. A. Chapman and G. A. Feldhammer, eds. Wild mammals of North America. Johns Hopkins Univ. Press, Baltimore.

Gates, J. E., and L. W. Gysel. 1978. Avian nest dispersion and fledging success in field-forest ecotones. Ecology 59:871–83.

Geer, G. I. 1997. Terrestrial salamanders at forest-farmland interfaces of Pennsylvania: Relative abundance, distribution, and microhabitat use. Master's thesis, Pennsylvania State Univ., University Park.

Geiger, R. 1965. The climate near the ground. Harvard Univ. Press, Cambridge.

Getz, L. L. 1961. Factors influencing the local distribution of shrews. Amer. Midland Nat. 65:67–88.

Gibbs, J. P. 1998. Amphibian movements in response to forest edges, roads, and streambeds in southern New England. J. Wildl. Manage. 62:584–89.
Gibbs, J. P., and J. Faaborg. 1990. Estimating the viability of ovenbird and Kentucky warbler populations in forest fragments. Conserv. Biol. 2:193–96.
Gibbs, J. P., and D. G. Wenny. 1993. Song output as a population estimator: Effect of male pairing success. J. Field Ornithol. 64:316–22.
Giese, R. L. 1990. Interactions of insects and forest trees. Pages 144–68 in R. A. Young and R. L. Giese, eds. Introduction to forest science. John Wiley & Sons, New York.
Gilbreath, J. 1998. 1997 marks a decade of success for the endangered red wolf. Red Wolf Newsletter 10(1):1–2.
Giles, R. H., Jr. 1978. Wildlife management. W. H. Freeman and Co., San Francisco.
Gill, F. B. 1990. Ornithology. W. H. Freeman and Co., New York.
Gilliam, F. S., N. L. Turrill, and M. B. Adams. 1995. Herbaceous-layer and overstory species in clear-cut and mature central Appalachian hardwood forests. Ecol. Appl. 5:947–55.
Gillis, A. M. 1990. The new forestry. BioScience 8:558–62.
Ginsberg, H. S. 1994. Lyme disease and conservation. Conserv. Biol. 8:343–53.
Giocomo, J. 1998. Effects of forest openings in the contiguous forest on the reproductive success of forest songbirds. M.S. thesis, Pennsylvania State Univ., University Park.
Gleason, H. A., and A. Cronquist. 1963. Manual of vascular plants of northeastern United States and adjacent Canada. D. Van Nostrand Co., New York.
Goodrich, L. J., and S. E. Senner. 1988. Recent trends of wintering great horned owls *(Bubo virginianus),* red-tailed hawks *(Buteo jamaicensis)* and two of their avian prey in Pennsylvania. J. Pennsylvania Acad. Sci. 62:131–37.
Graveland, J., R. van der Wal, J. H. van Balen, and A. J. van Noordwijk. 1994. Poor reproduction in forest passerines from decline of snail abundance on acidified soils. Nature 368:446–48.
Greenberg, C. H., and A. McGrane. 1996. A comparison of relative abundance and biomass of ground-dwelling arthropods under different forest management practices. For. Ecol. Manage. 89:31–41.
Grier, J. W., and R. W. Fyfe. 1987. Preventing research and management disturbance. Pages 173–213 in B. A. Giron Pendleton, B. A. Millsap, K. W. Cline, and D. M. Bird, eds. Raptor management techniques manual. National Wildlife Federation, Washington, D.C.
Grizzell, R. A., Jr. 1955. A study of the southern woodchuck, *Marmota monax monax.* Amer. Midland Nat. 53:257–93.
Gross, D. A. 1992. Rufous-sided towhee. Pages 512–15 in D. W Brauning, ed. Atlas of breeding birds in Pennsylvania. Univ. of Pittsburgh Press, Pittsburgh.
Grumbine, R. E. 1990. Viable populations, reserve size, and federal land management: A critique. Conserv. Biol. 4:127–34.

———. 1994. What is ecosystem management? Conserv. Biol. 8:27–38.
———. 1997. Reflections on "what is ecosystem management?" Conserv. Biol. 11:41–47.
Guildin, J. M. 1990. Multiple-use management, planning, and administration. Pages 197–208 in R. A. Young and R. L. Giese, eds. Introduction to forest science. John Wiley & Sons, New York.
Gullion, G. W. 1972. Improving your forested lands for ruffed grouse. Minnesota Agric. Exp. Stat., St Paul. Misc. J. Ser. Publ. 1439.
———. 1976. Ruffed grouse habitat manipulation—Mille Lacs Wildlife Management Area, Minnesota. Minnesota Wildl. Res. Quart. 36:97–121.
———. 1990. Forest-wildlife interactions. Pages 349–83 in R. A. Young and R. L. Giese, eds. Introduction to forest science. John Wiley & Sons, New York.
Gunson, J. R., and R. R. Bjorge. 1979. Winter denning of the striped skunk in Alberta. Can. Field-Nat. 93:252–58.
Gutzwiller, K. J. 1991. Assessing recreational impacts on wildlife: The value and design of experiments. Trans. N. Amer. Wildl. Nat. Resourc. Conf. 56:248–55.
Haas, C. A. 1995. Dispersal and use of corridors by birds in wooded patches on an agricultural landscape. Conserv. Biol. 9:845–54.
Hacker, D., and J. Renfro. 1992. Great Smoky Mountain plants studied for ozone sensitivity. Park Sci. 12(1):6–8.
Hafernik, J. E., Jr. 1992. Threats to invertebrate biodiversity: Implications for conservation strategies. Pages 171–95 in P. L. Fiedler and S. K. Jain, eds. Conservation biology. Chapman & Hall, New York.
Hagan, J. M. 1992. Conservation biology when there is no crisis—yet. Conserv. Biol. 6:475–76.
Hagan, J. M., III. 1993. Decline of the rufous-sided towhee in the eastern United States. Auk 110:863–74.
Hagan, J. M., W. M. Vander Haegen, and P. S. McKinley. 1996. The early development of forest fragmentation effects on birds. Conserv. Biol. 10:188–202.
Hagenstein, P. 1990. Forests. Pages 78–100 in R. N. Sampson and D. Hair, eds. Natural resources for the 21st century. Amer. For. Assoc., Island Press, Washington, D.C.
Hahn, D. C., and J. S. Hatfield. 1995. Parasitism at the landscape scale: Cowbirds prefer forest. Conserv. Biol. 9:1415–24.
Hammitt, W E., J. N. Dulin, and G. R. Wells. 1993. Determinants of quality wildlife viewing in Great Smoky Mountains National Park. Wildl. Soc. Bull. 21:21–30.
Harley, J. L. 1969. The biology of mycorrhizae. Leonard Hill, London.
Harrington, W. 1991. Wildlife: Severe decline and partial recovery. Pages 205–46 in K. D. Frederick and R. A. Sedjo, eds. America's renewable

resources: Historical trends and current challenges. Resources for the Future, Washington, D.C.

Harris, L. D. 1984. The fragmented forest. The Univ. of Chicago Press, Chicago.

Harris, L. D., and G. Silva-Lopez. 1992. Forest fragmentation and the conservation of biological diversity. Pages 197–237 in P. L. Fiedler and S. K. Jain, eds. Conservation biology. Chapman & Hall, New York.

Harris, L. D., and P. B. Gallagher. 1989. New initiatives for wildlife conservation: The need for movement corridors. Pages 11–34 in G. Mackintosh, ed. Preserving communities and corridors. Defenders of Wildlife, Washington, D.C.

Harrison, D. J., J. A. Bissonette, and J. A. Sherburne. 1989. Spatial relationships between coyotes and red foxes in eastern Maine. J. Wildl. Manage. 53:181–85.

Harrison, R. L. 1992. Toward a theory of inter-refuge corridor design. Conserv. Biol. 6:293–95.

Haskell, D. G. 1995. A reevaluation of the effects of forest fragmentation on rates of bird-nest predation. Conserv. Biol. 9:1316–18.

Heissenbuttel, J., and W. R. Murray. 1992. ESA: A troubled law in need of revision. J. Forestry 90(8):13–16.

Heinen, J. T. 1995. Thoughts and theory on incentive-based endangered species conservation in the United States. Wildl. Soc. Bull. 23:338–45.

Henein, K., and G. Merriam. 1990. The elements of connectivity where corridor quality is variable. Landscape Ecol. 4:157–70.

Henny, C. J., W. S. Seegar, and T. L. Maechtle. 1996. DDE decreases in plasma of spring migrant peregrine falcons, 1978–94. J. Wildl. Manage. 60:342–49.

Henry, J. D., and J. M. A. Swan. 1974. Reconstructing forest history from live and dead plant material—An approach to the study of forest succession in southwest New Hampshire. Ecology 55:772–83.

Heske, E. J. 1995. Mammalian abundances on forest-farm edges versus forest interiors in southern Illinois: Is there an edge effect? J. Mammal. 76:562–68.

Hess, G. R. 1994. Conservation corridors and contagious disease: A cautionary note. Conserv. Biol. 8:256–62.

Hesselton, W. T., and R. M. Hesselton. 1982. White-tailed deer. Pages 878–901 in J. A. Chapman and G. A. Feldhammer, eds. Wild mammals of North America. Johns Hopkins Univ. Press, Baltimore.

Hesser, R., R. Hoopes, C. B. Weirich, J. Selcher, B. Hollender, and R. Snyder. 1975. Clear-cutting in Pennsylvania: The aquatic biota. Pages 9–20 in Clear-cutting in Pennsylvania. School of Forest Resources, Pennsylvania State Univ., University Park.

Heusmann, H. W 1991. The history and status of the mallard in the Atlantic flyway. Wildl. Soc. Bull. 19:14–22.

Hickey, H. B., and M. C. Brittingham. 1991. Population dynamics of blue jays at a bird feeder. Wilson Bull. 103:401–14.

Hildebrand, E. S. 1994. Incidence of ozone-induced foliar injury on sensitive hardwood species in Shenandoah National Park, Virginia, 1991–92. Master's thesis, Pennsylvania State Univ., University Park.

Hildebrand, E., J. M. Skelly, and T. S. Fredericksen. 1996. Foliar response of ozone-sensitive hardwood tree species from 1991 to 1993 in the Shenandoah National Park, Virginia. Can. J. For. Res. 26:658–69.

Hill, E. P. 1982. Beaver. Pages 256–81 in J. A. Chapman and G. A. Feldhammer, eds. Wild mammals of North America. Johns Hopkins Univ. Press, Baltimore.

Hill, L. W. 1992. Ancient forests. J. Forestry 90(8):43.

Hirth, D. H. 1977. Social behavior of white-tailed deer in relation to habitat. Wildl. Monogr. 53:1–55.

Hobbs, N. T. 1996. Modification of ecosystems by ungulates. J. Wildl. Manage. 60:695–713.

Hobbs, R. J., and S. E. Humphries. 1995. An integrated approach to the ecology and management of plant invasions. Conserv. Biol. 9:761–70.

Hocker, H. W., Jr. 1979. Introduction to forest biology. John Wiley & Sons, New York.

Hodges, M. F., Jr., and D. G. Krementz. 1996. Neotropical migratory breeding bird communities in riparian forests of different widths along the Altamaha River, Georgia. Wilson Bull. 108:496–506.

Hollingsworth, R. W. 1988. Fish habitat and forest fragmentation. Pages 19–22 in R. M. DeGraaf and W. M. Healy, comps. Is forest fragmentation a management issue in the Northeast? USDA–For. Serv., Gen. Tech. Rept. NE-140. Radnor, Pa.

Hoover, J. P., and M. C. Brittingham. 1993. Regional variation in cowbird parasitism of wood thrushes. Wilson Bull. 105:228–38.

Hoover, J. P., M. C. Brittingham, and L. J. Goodrich. 1995. Effects of forest patch size on nesting success of wood thrushes. Auk 112:146–55.

Hornocker, M. G. 1970. An analysis of mountain lion predation upon mule deer and elk in the Idaho Primitive Area. Wildl. Monogr. 21:1–39.

Horsley, S. 1977. Allelopathic inhibition of black cherry by fern, grass, goldenrod, and aster. Can. J. For. Res. 7:205–16.

Horsley, S. B., and D. A. Marquis. 1983. Interference by weeds and deer with Allegheny hardwood reproduction. Can. J. For. Res. 13:61–69.

Houston, D. B., and E. G. Schreiner. 1995. Alien species in national parks: Drawing lines in space and time. Conserv. Biol. 9:204–09.

Houston, D. R. 1981. Effects of defoliation on trees and stands. Pages 217–97 in C. C. Doane and M. L. McManus, eds. The gypsy moth: Research toward integrated pest management. Tech. Bull. 1584, USDA–For. Serv., Washington, D.C.

Howard, W. E., R. L. Fenner, and H. E. Childs, Jr. 1959. Wildlife survival in brush burns. J. Range Manage. 12:530–34.

Howe, T. D., F. J. Singer, and B. A. Ackerman. 1981. Forage relationships of European wild boar invading northern hardwood forest. J. Wildl. Manage. 45:748–54.

Hughes, J. W., and T. J. Fahey. 1991. Availability, quantity, and selection of browse by white-tailed deer after clear-cutting. J. Forestry 89(10):31–36.

Humphrey, S. R. 1982. Bats. Pages 52–70 in J. A. Chapman and G. A. Feldhammer, eds. Wild mammals of North America. Johns Hopkins Univ. Press, Baltimore.

Hunt, F. A., and W. R. Irwin. 1992. A tough law to solve tough problems. J. Forestry 90(8):17–21.

Hunt, L. O. 1991. Forestry world games: "Biodiversity." J. Forestry 89(6):39.

Hunter, M. L., Jr. 1990. Wildlife, forests, and forestry. Prentice Hall, Englewood Cliffs, N.J.

Ingold, D. J. 1994. Influence of nest-site competition between European starlings and woodpeckers. Wilson Bull. 106:227–41.

Irwin, L. L., and T. B. Wigley. 1992. Conservation of endangered species—the impact on private forestry. J. Forestry 90(8):27–30, 42.

———. 1993. Toward an experimental basis for protecting forest wildlife. Ecol. Applications 3:213–17.

Jacobson, E., J. W. Carpenter, and M. Novilla. 1977. Suspected lead toxicosis in a bald eagle. J. Amer. Vet. Med. Assoc. 171:952–54.

James, F. C., C. E. McCullough, and D. A. Wiedenfeld. 1996. New approaches to the analysis of population trends in land birds. Ecology 77:13–27.

Jeffords, M. R., and A. G. Endress. 1984. A possible role of ozone in tree defoliation by the gypsy moth (Lepidoptera: Lymantridae). Envir. Entomol. 13:1249–52.

Johnson, A. S., W. M. Ford, and P. E. Hale. 1993. The effects of clear-cutting on herbaceous understories are still not fully known. Conserv. Biol. 7:433–35.

Johnson, G., and D. J. Leopold. 1998. Habitat management for the eastern massasauga rattlesnake in a central New York peatland. J. Wildl. Manage. 62:84–97.

Johnson, K. G., and M. R. Pelton. 1981. Selection and availability of dens for black bears in Tennessee. J. Wildl. Manage. 45:111–19.

Johnson, W. C., and C. S. Adkisson. 1986. Airlifting the oaks. Nat. Hist. 95:40–47.

Kamalay, J. C. 1996. Hope for DED. Arbor Age, July:50.

Kapos, V. 1989. Effects of isolation on the water status of forest patches in the Brazilian Amazon. J. Tropical Ecol. 5:173–85.

Karr, J. R. 1990. Biological integrity and the goal of environmental legislation: Lessons for conservation biology. Conserv. Biol. 4:244–50.

Kasbohm, J. W., M. R. Vaughan, and J. G. Kraus. 1995. Food habits and nutrition of black bears during a gypsy moth infestation. Can. J. Zool. 73:1771–75.

———. 1996. Effects of gypsy moth infestation on black bear reproduction and survival. J. Wildl. Manage. 60:408–416.

Kattan, G. H., H. Alvarez-López, and M. Giraldo. 1994. Forest fragmentation and bird extinctions: San Antonio eighty years later. Conserv. Biol. 8:138–46.

Keddy, P. A., and C. G. Drummond. 1996. Ecological properties for the evaluation, management, and restoration of temperate deciduous forest ecosystem. Ecol. Applications 6:748–62.

Keith, L. B., J. R. Cary, O. J. Rongstad, and M. C. Brittingham. 1984. Demography and ecology of a declining snowshoe hare population. Wildl. Monogr. 90:1–43.

Kellert, S. R. 1985. Bird-watching in American society. Leisure Sci. 7:343–60.

Kellert, S. R. 1995. Managing for biological and sociological diversity, or 'deja vu, all over again.' Wildl. Soc. Bull. 23:274–78.

Kennedy, E. D., and D. W. White. 1996. Interference competition from house wrens as a factor in the decline of Bewick's wrens. Conserv. Biol. 10:281–84.

Kerr, R. A. 1996. Ozone-destroying chlorine tops out. Science 271:32.

Keystone Center. 1991. Biological diversity on federal lands. The Keystone Center, Keystone, Colo.

Kiester, A. R., J. M. Scott, B. Csuti, R. F. Noss, B. Butterfield, K. Sahr, and D. White. 1996. Conservation prioritization using GAP data. Conserv. Biol. 10:1332–42.

Kilgo, J. C., R. A. Sargeant, K. V. Miller, and B. R. Chapman. 1997. Landscape influences on breeding bird communities in hardwood fragments of South Carolina. Wildl. Soc. Bull. 25:878–85.

Kilgo, J. C., R. A. Sargent, B. R. Chapman, and K. V. Miller. 1998. Effect of stand width and adjacent habitat on breeding bird communities in bottomland hardwoods. J. Wildl. Manage. 62:72–83.

Kiltie, R. A. 1989. Wildfire and the evolution of dorsal melanism in fox squirrels, *Sciurus niger.* J. Mammal. 70:726–39.

Kim, K. C. 1993. Biodiversity, conservation and inventory: Why insects matter. Biodiversity Conserv. 2:191–214.

Kimmel, J. T., and R. H. Yahner. 1994. The northern goshawk in Pennsylvania: Habitat use, survey protocols, and status. Final Rept., School of Forest Resources, Pennsylvania State Univ., University Park.

Kimmerer, T. W. 1990. Structure and function of forest trees. Pages 67–85 in R. A. Young and R. L. Giese, eds. Introduction to forest science. John Wiley & Sons, New York.

King, D. I., C. R. Griffin, and R. M. DeGraaf. 1996. Effects of clear-cutting on habitat use and reproductive success of the ovenbird in forested landscapes. Conserv. Biol. 10:1380–86.

Kirkland, G. L., Jr., A. F. Rhoads, and K. C. Kim. 1990. Perspectives on biodiversity in Pennsylvania and its maintenance. J. Pennsylvania Acad. Sci. 64:155–59.

Kirkpatrick, J. F., and J. W. Turner, Jr. 1985. Chemical fertility control and wildlife management. BioScience 35:485–91.

———. 1995. Urban deer fertility control: Scientific, social and political issues. Northeast Wildl. 52:103–16.

Kirkpatrick, J. F., I. K. M. Liu, J. W. Turner, Jr., R. Naugle, and R. Keiper. 1992. Long-term effects of porcine zonae pellucidae immunocontraception on ovarian function of feral horses *(Equus caballus).* J. Reprod. Fert. 94:437–44.

Kitchings, J. T, and B. T. Walton. 1991. Fauna of the North American temperate deciduous forest. Pages 345–70 in D. W. Goodall, ed. Temperate ecosystems of the worlds. Elsevier, New York.

Knight, R. L. 1996. Aldo Leopold, the land ethic, and ecosystem management. J. Wildl. Manage. 60:471–74.

Knight, R. L., and G. K. Meffe. 1997. Ecosystem management: Agency liberation from command and control. Wildl. Soc. Bull. 25:676–78.

Kolb, T. E., L. H. McCormick, E. E. Simons, and D. J. Jeffrey. 1992. Impacts of pear thrips on root carbohydrate, sap and crown characteristics of sugar maples in a Pennsylvania sugarbush. For. Sci. 38:381–92.

Komarek, E. V., Jr. 1969. Fire and animal behavior. Tall Timbers Fire Ecol. Conf. 9:161–207.

Kozicky, E. L. 1943. Food habits of foxes in wild turkey territory. Pennsylvania Game News 14:8–9, 28.

Kozlowski, T. T. 1971. Growth and development. 1. Seed germination, ontogeny, and short growth. Academic Press, New York.

Krebs, C. J. 1972. Ecology: The experimental analysis of distribution and abundance. Harper & Row, New York.

———. 1996. Population cycles revisited. J. Mammal. 77:8–24.

Krebs, J. W., M. L. Wilson, and J. E. Childs. 1995. Rabies—epidemiology, prevention, and future research. J. Mammal. 76:681–94.

Krementz, D. G., J. T. Seginak, and G. W. Pendleton. 1994. Winter movements and spring migration of American woodcock along the Atlantic coast. Wilson Bull. 106:482–93.

Kunz, T. H. 1982. Ecology of bats. Plenum Publ. Corp., New York.

Kurz, W. A., and R. N. Sampson. 1991. North American forests and global climatic change. Trans. N. Amer. Wildl. Nat. Resourc. Conf. 56:587–94.

Kuusipalo, J., and J. Kangas. 1994. Managing biodiversity in a forestry environment. Conserv. Biol. 8:450–60.

Laarman, J. G., and R. A. Sedjo. 1992. Global forests: Issues for six billion people. McGraw-Hill, New York.

Lacasse, N. L. 1994. Capital Area Greenbelt master plan announced. Pennsylvania Forests 85(2):7.

Lacher, T. E., Jr., and M. A. Mares. 1996. Availability of resources and use of space in eastern chipmunks, *Tamias striatus.* J. Mammal. 77:833–49.
Laurance, W. F., K. R. McDonald, and R. Speare. 1996. Epidemic disease and the catastrophic decline in Australian rain tree frogs. Conserv. Biol. 10:406–13.
Lautenschlager, R. A. 1997. Biodiversity is dead. Wildl. Soc. Bull. 25:679–35.
Ledig, F. T. 1992. Human impacts on genetic diversity in forest ecosystems. Oikos 63:87–108.
Ledig, F. T., and J. H. Kitzmiller. 1992. Genetic strategies for reforestation in the face of global climate change. For. Ecol. Manage. 50:153–69.
LeDuc, J. 1987. Epidemiology of Hantaan and related viruses. Laboratory Anim. Sci. 37:413–18.
Leimgruber, P., W. J. McShea, and J. R. Rappole. 1994. Predation of artificial nests in large forest blocks. J. Wildl. Manage. 58:254–60.
Leonard, D. E. 1981. Bioecology of the gypsy moth. Pages 9–29 in C. C. Doane and M. L. McManus, eds. The gypsy moth: Research toward integrated pest management. Tech. Bull. 1584, USDA–For. Serv., Washington, D.C.
Leopold, A. 1933. Game management. Charles Scribner's Sons, New York.
———. 1949. A Sand County almanac. Oxford Univ. Press, New York.
Levin, D. A., J. Francisco-Ortega, and R. K. Jansen. 1996. Hybridization and the extinction of rare plant species. Conserv. Biol. 10:10–16.
Levin, S. A. 1993. Science and sustainability. Ecol. Applications 3:545–46.
Lewis, A. R. 1982. Selection of nuts by gray squirrels and optimal foraging theory. Amer. Midland Nat. 107:250–57.
Lewis, A. R., and R. H. Yahner. 1999. Sex-specific habitat use by eastern towhees in a managed forested landscape. Jour. Penn. Acad. Science 16:1–7.
Lidicker, W. Z., Jr. 1988. Solving the enigma of microtine "cycles." J. Mammal. 69:225–35.
Lima, S. L., and P. A. Zollner. 1996. Towards a behavioral ecology of ecological landscapes. TREE 11:131–34.
Liebhold, A. M., J. A. Halverson, and G. A. Elmes. 1992. Gypsy moth invasion in North America: A quantitative analysis. J. Biogeography 19:513–20.
Lindemayer, D. B., and H. A. Nix. 1993. Ecological principles for the design of wildlife corridors. Conserv. Biol. 7:627–30.
Lindström, E. R., H. Andrén, P. Angelstam, G. Cederlund, B. Hörnfeldt, L. Jäderburg, P.-A. Lemnell, B. Martinsson, K. Sköld, and J. E. Swenson. 1994. Disease reveals the predator: Sarcoptic mange, red fox predation, and prey populations. Ecology 75:1042–49.
Lindzey, F. G., and E. C. Meslow. 1976. Winter dormancy in black bears in southwestern Washington. J. Wildl. Manage. 40:408–15.
Little, E. L., Jr. 1971. Atlas of United States trees. Vol. 1, Conifers and important hardwoods. Misc. Publ. 1146, USDA–For. Serv., Washington, D.C.
———. 1979. Checklist of United States trees (native and naturalized). Agric. Handbk. 541, USDA–For. Serv., Washington, D.C.

Litvaitis, J. A. 1993. Response of early successional vertebrates to historic changes in land use. Conserv. Biol. 7:866–81.

Litvaitis, J. A., and R. Villafuerte. 1996. Factors affecting the persistence of New England cottontail metapopulations: The role of habitat management. Wildl. Soc. Bull. 24:686–93.

Lorimer, C. G. 1985. The role of fire in the perpetuation of oak forest. Pages 8–25 in J. E. Johnson, ed. Challenges in oak management and utilization. Coop. Extension Serv., University of Wisconsin-Madison.

———. 1990a. Silviculture. Pages 300–25 in R. A. Young and R. L. Giese, eds. Introduction to forest science. John Wiley & Sons, New York.

———. 1990b. Behavior and management of forest fires. Pages 427–48 in R. A. Young and R. L. Giese, eds. Introduction to forest science. John Wiley & Sons, New York.

Lorimer, C. G., and L. E. Frelich. 1994. Natural disturbance regimes in old-growth northern hardwoods: Implications for restoration efforts. J. Forestry 92(1):33–38.

Lubchenco, J. A., M. Olson, L. B. Brubaker, S. R. Carpenter, M. M. Holland, S. P. Hubbell, S. A. Levin, J. A. MacMahon, P. A. Matson, J. M. Melillo, H. A. Mooney, C. H. Peterson, H. R. Pulliam, L. A. Real, P. J. Regal, and P. G. Risser. 1991. The Sustainable Biosphere Initiative: An ecological research agenda. Ecology 72:371–412.

Lynch, J. A., K. S. Horner, J. W. Grimm, and E. S. Corbett. 1993. Atmospheric deposition: Spatial and temporal variations in Pennsylvania—1992. Environ. Resourc. Res. Instit., Pennsylvania State Univ., University Park.

Lynch, J. F., and D. F. Whigham. 1984. Effects of forest fragmentation on breeding bird communities in Maryland, USA. Biol. Conserv. 28:287–324.

MacArthur, R. H., and E. O. Wilson. 1967. The theory of island biogeography. Princeton Univ. Press, Princeton, N.J.

MacCleery, D. W. 1992. American forests: A history of resiliency and recovery. USDA–For. Serv., Washington, D.C. FS-540.

Machtans, C. S., M.-A. Villard, and S. J. Hannon. 1996. Use of riparian buffer strips as movement corridors by forest birds. Conserv. Biol. 10:1366–79.

Maehr, D. S., E. D. Land, and M. E. Roelke. 1991. Mortality patterns of panthers in southwest Florida. Proc. Ann. Conf. Southeast. Assoc. Fish Wildl. Agencies 45:201–07.

Mahan, C. G. 1996. The ecology of eastern chipmunks *(Tamias striatus)* in a fragmented forest. Ph.D. dissertation, Pennsylvania State Univ., University Park.

Mahan, C. G., K. A. Sullivan, K. C. Kim, R. H. Yahner, and M. D. Abrams. 1997. Biodiversity profile assessment of eastern hemlock forests: Sampling protocols and procedures. Draft Final Rept., USDI, Nat. Park Serv., Mid-Atlantic Region, Philadelphia, Pa.

Mahan, C. G., and R. H. Yahner. 1992. Microhabitat use by red squirrels in central Pennsylvania. Trans. Northeast Sect. Wildl. Soc. 49:49–56.

———. 1996. Effects of forest fragmentation on burrow-site selection by the eastern chipmunk *(Tamias striatus).* Amer. Midland Nat. 136:352–57.

Mahlman, J. D. 1992. A looming Arctic ozone hole? Nature 360:209–10.

Manfredo, M. J., M. Fishbein, G. E. Haas, and A. E. Watson. 1990. Attitudes toward prescribed fire policies. J. Forestry 88(7):19–23.

Marquis, D. A. 1975. The Allegheny hardwood forests of Pennsylvania. USDA–For. Serv., Northeastern For. Exp. Stat., Gen. Tech. Rept. NE-15. Upper Darby, Pa.

———. 1981. Effect of deer browsing on timber production in Allegheny hardwood forests of northwestern Pennsylvania. USDA–For. Serv. Res. Pap. NE-475, Broomall, Pa.

———. 1983. Ecological and historical background: Northern hardwoods. Pages 9–29 in J. Finley, R. S. Cochran, and J. R. Grace, eds. Proc. of regenerating hardwood stands. Penn State Forestry Issues Conf., Pennsylvania State Univ., University Park.

Marquis, D. A., and T. J. Grisez. 1978. The effect of deer exclosures on the recovery of vegetation in failed clear-cuts on the Allegheny Plateau. USDA–For. Serv. Res. Note NE-270, Broomall, Pa.

Marquis, R. J., and C. J. Whelan. 1994. Insectivorous birds increase growth of white oak through consumption of leaf-eating insects. Ecology 75:2007–14.

Marti, C. D., and M. N. Kochert. 1995. Are red-tailed hawks and great horned owls diurnal-noctural dietary counterparts? Wilson Bull. 107:615–28.

Martin, A. C., H. S. Zimm, and A. L. Nelson. 1951. American wildlife and plants: A guide to wildlife food habits. McGraw-Hill, New York.

Martin, A. J., and J. Bliss. 1990. Nonindustrial private forests. Pages 231–53 in R. A. Young and R. L. Giese, eds. Introduction to forest science. John Wiley & Sons, New York.

Martin, C. 1993. Partners in Flight—Aves de las Americas. Fish and Wildlife News, USDA. Winter:11.

Martin, T. E. 1980. Diversity and abundance of spring migratory birds using habitat islands on the Great Plains. Condor 82:430–39.

Marx, D. H., and D. J. Beattie. 1977. Mycorrhizae—promising aid to timber farmers. For. Farmer 36:1–9.

Masters, R. E. 1991. Effects of fire and timber harvest on vegetation and cervid use of oak-pine sites in Oklahoma Ouachita Mountains. Pages 168–76 in S. C. Nodvin and T. A. Waldrop, eds. Fire and the environment: Ecological and cultural perspectives. Southeastern For. Exp. Stat., Asheville, N. C. Gen. Tech. Rept. SE-69.

Mastrota, F. N., R. H. Yahner, and G. L. Storm. 1989. Small mammal communities in a mixed-oak forest irrigated with wastewater. Amer. Midland Nat. 122:388–93.

Mather, A. S. 1990. Global forest resources. Timber Press, Portland, Oreg.

Mattfield, G. F. 1984. Pages 305–30 in L. K. Halls, ed. White-tailed deer: Ecology and management. Stackpole Books, Harrisburg, Pa.

Matthews, J. R., originating ed. 1991. The official World Wildlife Fund guide to endangered species of North America. Vol. 1. Beacham Publishing, Washington, D.C.

Matthiessen, P. 1978. Wildlife in America. Penguin Books, New York.

Mattson, D. J., and M. M. Reid. 1991. Conservation of the Yellowstone grizzly bear. Conserv. Biol. 5:364–72.

Mattson, W. J., and N. D. Addy. 1975. Phytophagous insects as regulators of forest primary production. Science 190:515–22.

Matuszek, J. R., and G. L. Beggs. 1988. Fish species richness in relation to lake area, pH, and other abiotic factors in Ontario lakes. Can. J. Fish. Aqua. Sci. 45:1931–41.

Maurer, B. A., and S. G. Heywood. 1993. Geographic range fragmentation and abundance of Neotropical migratory birds. Conserv. Biol. 7:501–09.

Mautz, W. W. 1984. Sledding on a bushy hillside: The fat cycle in deer. Wildl. Soc. Bull. 6:88–90.

Mayfield, H. 1965. The brown-headed cowbird, with old and new hosts. Living Bird 4:13–28.

———. 1977. Brown-headed cowbird: Agent of extermination? Amer. Birds 31:107–13.

McAninch, J. B., and J. M. Parker. 1991. Urban deer management programs: A facilitated approach. Trans. N. Amer. Wildl. Nat. Res. Conf. 56:428–35.

McAuley, D. G., and J. R. Longcore. 1988. Survival of juvenile ring-necked ducks on wetlands of different pH. J. Wildl. Manage. 52:169–76.

McCabe, R. E., and T. R. McCabe. 1984. Of slings and arrows: An historical retrospection. Pages 19–72 in L. K. Halls, ed. White-tailed deer: Ecology and management. Stackpole Books, Harrisburg, Pa.

McClanahan, T. R., and R. W. Wolfe. 1993. Accelerating forest succession in a fragmented landscape: The role of birds and perches. Conserv. Biol. 7:279–88.

McClenahen, J. R., R. J. Hutnik, and D. D. Davis. 1997. Patterns of northern red oak growth and mortality in western Pennsylvania. Pages 386–99 in S. G. Pallardy, R. A. Cecich, H. G. Garrett, and P. S. Johnson, eds. Proceedings Central Hardwood Forest Conf., Columbia, Mo.

McClure, M. S., S. M. Salom, and K. S. Shields. 1996. Hemlock woolly adelgid. USDA–For. Serv., FHTET-96-35, Morgantown, W.Va.

McComb, W. C., and R. Noble. 1981. Nest-box and natural-cavity use in three mid-South forest habitats. J. Wildl. Manage. 45:93–101.

McDonald, J. E., Jr., W. L. Palmer, G. L. Storm. 1994. Ruffed grouse population response to intensive forest management in central Pennsylvania, USA. Proc. Internat. Union Game Biol. Congress 21:126–31.

McDonald, K. A., and J. H. Brown. 1992. Using montane mammals to model extinctions due to global change. Conserv. Biol. 6:409–15.

McIntyre, S., and G. W. Barrett. 1992. Habitat variegation, an alternative to fragmentation. Conserv. Biol. 6:146–47.

McLean, R. B. 1970. Wildlife rabies in the United States: Recent history and current concepts. J. Wildl. Diseases 6:229–35.

McMinn, J. W 1991. Biological diversity research: An analysis. USDA–For. Serv., Gen. Tech. Rept. SE-71.

McShea, W. J., and J. H. Rappole. 1997. Variable song rates in three species of passerines and implications for estimating bird populations. J. Field Ornithol. 68:367–75.

McShea, W. J., and G. Schwede. 1993. Variable acorn crops: Responses of white-tailed deer and other consumers. J. Mammal. 74:999–1006.

Means, D. B., J. G. Pallis, and M. Baggett. 1996. Effects of slash pine silviculture on a Florida population of flatwoods salamander. Conserv. Biol. 10:426–37.

Mech, L. D. 1977. Wolf-pack buffer zones as prey reservoirs. Science 198:320–21.

———. 1994. Buffer zones of territories of gray wolves as regions of intraspecific strife. J. Mammal. 75:199–202.

Meier, A. J., S. P. Bratton, and D. C. Duffy. 1995. Possible ecological mechanisms for loss of vernal-herb diversity in logged eastern deciduous forests. Ecol. Appl. 5:935–46.

Menges, E. S. 1991. Seed germination percentage increases with population size in a fragmented prairie species. Conserv. Biol. 5:158–64.

Merriam, G., and A. Lanoue. 1990. Corridor use by small mammals: Field measurement for three experimental types of *Peromyscus leucopus.* Landscape Ecol. 4:123–31.

Merrill, S. B., F. J. Cuthbert, and G. Oehlert. 1998. Residual patches and their contribution to forest-bird diversity on northern Minnesota aspen clear-cuts. Conserv. Biol. 12:190–99.

Merritt, J. F. 1987. Guide to the mammals of Pennsylvania. Univ. of Pittsburgh Press, Pittsburgh.

Meyer, J. M. 1992. Rethinking the outlook of colleges whose roots have been in agriculture. Univ. of California, Davis.

Miller, G. 1996. Ecosystem management: Improving the Endangered Species Act. Ecol. Applications 6:715–77.

Miller, G. T., Jr. 1990. Resource conservation and management. Wadsworth, Belmont, Calif.

———. 1992. Living in the environment. 7th ed. Wadsworth, Belmont, Calif.

Miller, G. W., J. E. Johnson, and J. E. Baumgras. 1997. Deferment cutting in central Appalachian hardwoods. For. Landowner 56(5):28–31, 68.

Miller, G. W., P. B. Wood, and J. V. Nichols. 1995. Two-age silviculture—an innovative tool for enhancing species diversity and vertical structure in Appalachian hardwoods. Pages 175–82 in L. G. Eskew, comp. Proc. of the national silviculture workshop on forest health through silviculture. USDA–For. Serv., Gen. Tech. Rept. RM-GRT-267, Fort Collins, Colo.

Mills, J. N., T. L. Yates, J. E. Childs, R. P. Parmenter, T. G. Ksiazek, P. E. Rollin, and C. J. Peters. 1995. Guidelines for working with rodents potentially infected with hantavirus. J. Mammal. 76:716–22.

Mills, K. H., and D. W. Schindler. 1986. Biological indicators of lake acidification. Water Air Soil Pollut. 30:779–89.

Mladenoff, D. J., and T. A. Sickley. 1998. Assessing potential gray wolf restoration in the northeastern United States: A spatial prediction of favorable habitat and potential population levels. J. Wildl. Manage. 61:1–10.

Mladenoff, D. J., and F. Stearns. 1993. Eastern hemlock regeneration and deer browsing in the northern Great Lakes region: A reexamination and model simulation. Conserv. Biol. 7:889–900.

Mlot, C. 1989. Great Lakes fish and the greenhouse effect. BioScience 39:145.

———. 1992. Botanists sue Forest Service to preserve biodiversity. Science 257:1618–19.

Moehring, D. M., C. X. Grano, and J. R. Bassett. 1966. Properties of forested loess soils after repeated prescribed burns. USDA–For. Serv., Res. Note SO-40. Southern For. Exp. Stat., New Orleans.

Montzka, S. A., J. H. Butler, R. C. Myers, T. M. Thompson, T. H. Swanson, A. D. Clarke, L. T. Lock, and J. W. Elkins. 1996. Decline in the tropospheric abundance of halogen form halocarbons: Implications for stratospheric ozone depletion. Science 272:1318–22.

Moore, F. R., and W. Yong. 1991. Evidence of food-based competition among passerine migrants during stopover. Behav. Ecol. Sociobiol. 28:85–90.

Moorman, C. E., and B. R. Chapman. 1996. Nest-site selection of red-shouldered and red-tailed hawks in a managed forest. Wilson Bull. 108:357–68.

Morrell, T. E., and R. H. Yahner. 1994. Habitat characteristics of great horned owls in south-central Pennsylvania. J. Raptor Res. 28:164–70.

Morris, S. R., M. E. Richmond, and D. W. Holmes. 1994. Patterns of stopover by warblers during spring and fall migration on Appledore Island, Maine. Wilson Bull. 106:703–18.

Morrison, M. L., I. C. Timossi, K. A. With, and P. N. Manley. 1985. Use of tree species by forest birds during winter and summer. J. Wildl. Manage. 49:1098–1102.

Moyle, P. B., and R. A. Leidy. 1992. Loss of biodiversity in aquatic ecosystems: Evidence from fish faunas. Pages 127–69 in P. L. Fiedler and S. K. Jain, eds. Conservation biology. Chapman & Hall, New York.

Moyle, P. B., and J. E. Williams. 1990. Biodiversity loss in the temperate zone: Decline of native fish fauna of California. Conserv. Biol. 4:275–84.

Muller, R. N. 1978. The phenology, growth, and ecosystem dynamics of *Erythronium americanum* in the northern hardwood forest. Ecol. Monogr. 48:1–20.

Muller, R. N., and F. H. Bormann. 1976. Role of *Erythronium americanum* Ker. in energy flow and nutrient dynamics of a northern hardwood ecosystem. Science 193:1126–28.

Muller, W. W. 1994. Oral vaccination of foxes in Europe, 1992. Rabies Bull. Europe 16:12–13.

Murphy, D. D. 1989. Conservation and confusion: Wrong species, wrong scale, wrong conclusions. Conserv. Biol. 3:82–84.

Murphy, D., D. Wilcove, R. Noss, J. Harte, C. Safina, J. Lubchenco, T. Root, V. Sher, L. Kaufman, M. Bean, and S. Pimm. 1994. On reauthorization of the Endangered Species Act. Conserv. Biol. 8:1–3.

Myers, N. 1988. Tropical forests and their species. Pages 28–35 in E. O. Wilson, ed. Biodiversity. National Academy Press, Washington, D.C.

———. 1993. Environmental refugees in a globally warmed world. BioScience 43:752–61.

Nash, B. L., D. D. Davis, and J. M. Skelly. 1992. Forest health along a wet sulfate/pH deposition gradient in north-central Pennsylvania. Environ. Toxicol. Chem. 11:1095–1104.

Nater, E. A., and D. F. Grigal. 1992. Regional trends in mercury distribution across the Great Lake states, north-central USA. Nature 358:139–41.

National Ecosystem Management Forum. 1993. Meeting summary. The Keystone Center, Keystone, Colo.

National Park Service. 1988. Management policies. National Park Service, Washington, D.C.

Nelson, B., C. Nowak, S. Reitz, D. deCalesta, and S. Wingate. 1997. Communicating old-growth forest management on the Allegheny National Forest. Pages 77–81 in Communicating the role of silviculture in managing the national forests. Proc. of the national silviculture workshop. USDA–For. Serv., Northeast. For. Exp. Stat., Radnor, Pa., Gen. Tech. Rep. NE-238.

Nelson, M. E., and L. D. Mech. 1981. Deer social organization and wolf predation in northeastern Minnesota. Wildl. Monogr. 77:1–53.

Neue, H.-V. 1993. Methane emission from rice fields. BioScience 43:466–74.

Newark, W. D. 1995. Extinction of mammal populations in western North American national parks. Conserv. Biol. 9:512–26.

Nicholls, M. S. 1992. Blame it on Rio? BioScience 42:652.

Nicholson, R. 1992. Death and Taxus. Natural Hist. 9/92:20–23.

Niering, W. A., and N. C. Olmstead. 1983. The Audubon Society field guide to North American wildflowers. Alfred A. Knopf, New York.

Nigh, T. A., W. L. Pflieger, P. L. Redfearn, W. A. Schroeder, A. R. Templeton, and F. R. Thompson III. 1992. The biodiversity of Missouri. U. S. Fish Wildl. Serv., Washington, D.C.

Nixon, C. M., S. P. Havera, and L. P. Hansen. 1980. Initial response of squirrels to forest changes associated with selection cutting. Wildl. Soc. Bull. 8:298–306.

Nixon, C. M., M. W. McClain, and R. W. Donohoe. 1975. Effects of hunting and mast crops on a squirrel population. J. Wildl. Manage. 39:1–25.

———. 1980. Effects of clear-cutting on gray squirrels. J. Wildl. Manage. 44:403–12.

Noss, R. F. 1983. A regional landscape approach to maintain diversity, BioScience 33:700–06.

———. 1987. Corridors in real landscapes: A reply to Simberloff and Cox. Conserv. Biol. 1:159–64.

———. 1991. Sustainability and wilderness. Conserv. Biol. 5:120–22.

———. 1992. Issues of scale in conservation biology. Pages 239–50 in P. L. Fiedler and S. K. Jain, eds. Conservation biology. Chapman & Hall, New York.

———. 1996. The naturalists are dying off. Conserv. Biol. 10:1–3.

Noss, R. F., and A. Y. Cooperrider. 1994. Saving nature's legacy: Protecting and restoring biodiversity. Island Press, Washington, D.C.

Nowacki, G. J., and M. D. Abrams. 1992. Community, edaphic, and historical analysis of mixed oak forests of the Ridge and Valley Province of central Pennsylvania. Can. J. For. Res. 22:790–800.

Nowak, D. J. 1994a. Air pollution removal by Chicago's urban forest. Pages 63–81 in E. G. McPherson, D. J. Nowak, and R. A. Roundtree, eds. Chicago's urban forest ecosystem: Results of the Chicago Urban Forest Climate Project. Gen. Tech. Rept. NE-186. USDA–For. Serv., Radnor, Pa.

———. 1994b. Atmospheric carbon dioxide reduction by Chicago's urban forest. Pages 83–94 in E. G. McPherson, D. J. Nowak, and R. A. Roundtree, eds. Chicago's urban forest ecosystem: Results of the Chicago Urban Forest Climate Project. Gen. Tech. Rept. NE-186. USDA–For. Serv., Radnor, Pa.

Nowak, R. M. 1995. Another look at wolf taxonomy. Pages 375–97 in L. N. Carbyn, S. H. Fritts, and D. R. Seip, eds. Ecology and conservation of wolves in a changing world. Canadian Circumpolar Institute, Edmonton, Alberta.

Nowak, R. M. 1999. Walker's mammals of the world. Vol. II. Sixth ed. The Johns Hopkins University Press, Baltimore.

O'Donnell, E. 1992. Pennsylvania forest stewardship. School of Forest Resources, Pennsylvania State Univ., University Park.

Oehler, J. D., and J. A. Litvaitis. 1996. The role of spatial scale in understanding responses of medium-sized carnivores to forest fragmentation. Can. J. Zool. 74:2070–79.

Office of Technology Assessment. 1993. Preparing for an uncertain climate. U.S. Congress, Washington, D.C. OTA-O-563.

Ogden, J. C. 1992. The impact of Hurricane Andrew on the ecosystem of south Florida. Conserv. Biol. 6:488–90.

Ojima, D. S., K. A. Galvin, and B. L. Turner II. 1994. The global impact of land-use change. BioScience 44:300–04.

O'Keefe, T. 1990. Holistic (new) forestry: Significant difference or just another gimmick? J. Forestry 90(10):41–44.
Ola, P., and E. D'Aulaire. 1997. Get ready for some wild weather. Reader's Digest, November:92–96.
Oli, M. K., H. A. Jacobson, and B. D. Leopold. 1997. Denning ecology of black bears in the White River National Wildlife Refuge, Arkansas. J. Wildl. Manage. 61:700–6.
Oliver, C. D. 1992. Achieving and maintaining biodiversity and economic productivity: A landscape approach. J. Forestry 90(9):20–25.
Oliver, C. D., and B. C. Larson. 1996. Forest stand dynamics. John Wiley & Sons, Inc., New York.
O'Meara, M. 1997. The risks of disrupting climate. World-Watch, November/December:10–35.
Onken, A. 1995. Gypsy moth nucleopolyhedrosis virus (NPV) literature review. Gypsy Moth News, October:7–10.
Orr, D. W. 1993. Forests and trees. Conserv. Biol. 7:454–56.
Ostfeld, R. S. 1997. The ecology of Lyme-disease risk. Amer. Sci. 85:338–46.
Ostfeld, R. S., M. C. Miller, and K. R. Hazler. 1996. Causes and consequences of tick *(Ixodes scapularis)* burdens on white-footed mice *(Peromyscus leucopus).* J. Mammal. 77:266–73.
Ostfeld, R. S., C. G. Jones, and J. O. Wolff. 1996. Of mice and mast. BioScience 46:323–30.
Ozoga, J. J., L. J. Verme, and C. S. Bienz. 1982. Parturition behavior and territoriality in white-tailed deer: Impact on neonatal mortality. J. Wildl. Manage. 46:1–11.
Palomares, F., P. Gaona, P. Ferreras, and M. Delibes. 1995. Positive effects of game species of top predators by controlling smaller predator populations: An example with lynx, mongooses, and rabbits. Conserv. Biol. 9:295–305.
Paradiso, J. L., and R. M. Nowak. 1982. Wolves. Pages 460–74 in J. A. Chapman and G. A. Feldhammer, eds. Wild mammals of North America. Johns Hopkins Univ. Press, Baltimore.
Pashley, D., C. Hunter, and M. Carter. 1992. Species prioritization scheme. Partners in Flight 2(3):1, 21–22.
Paton, W. C. 1994. The effect of edge on avian nest success: How strong is the evidence? Conserv. Biol. 8:17–26.
Patton, D. R. 1992. Wildlife habitat relationships in forested ecosystems. Timber Press, Portland, Oreg.
Patton, R. F. 1990. Diseases of forest trees. Pages 169–94 in R. A. Young and R. L. Giese, eds. Introduction to forest science. John Wiley & Sons, New York.
Peek, J. M. 1980. Natural regulation of ungulates (what constitutes a real wilderness?). Wildl. Soc. Bull. 8:217–27.
Pelton, M. R. 1982. Black bear *(Ursus americanus).* Pages 504–14 in J. A.

Chapman and G. A. Feldhamer, eds. Wild mammals of North America. Johns Hopkins Univ. Press, Baltimore, Md.

Pennsylvania Bureau of Forestry. 1991. Silvicultural modifications for biodiversity—reservation guidelines. Pennsylvania Bureau of Forestry, Harrisburg.

Perlin, J. 1989. A forest journey: The role of wood in the development of civilization. Harvard Univ. Press, Cambridge.

Perry, D. A. 1993. Biodiversity and wildlife are not synonymous. Conserv. Biol. 7:204–5.

Peterjohn, B. G., and J. R. Sauer. 1994. Population trends of woodland birds from the North American Breeding Bird Survey. Wildl. Soc. Bull. 22:155–64.

Peters, R. L., II. 1988. The effect of global climatic change on natural communities. Pages 450–61 in E. O. Wilson, ed. Biodiversity. National Academy Press, Washington, D.C.

Peters, R. P., and L. D. Mech. 1975. Scent-marking in wolves. Amer. Scientist 63:628–37.

Peterson, R. O., and R. E. Page. 1988. The rise and fall of the Isle Royale wolves, 1975–1986. J. Mammal. 69:89–99.

Petit, D. R., K. E. Petit, and T. C. Grubb Jr. 1985. On atmospheric moisture as a factor influencing distribution of breeding birds in temperate deciduous forest. Wilson Bull. 97:88–96.

Petit, L. J., D. R. Petit, and T. E. Martin. 1995. Landscape-level management of migratory birds: Looking past the trees to see the forest. Wildl. Soc. Bull. 23:420–29.

Petranka, J. W., M. E. Eldridge, and K. E. Haley. 1993. Effects of timber harvesting on southern Appalachian salamanders. Conserv. Biol. 7:363–70.

Phillips, M. K., and W. T. Parker. 1988. Red wolf recovery: A progress report. Conserv. Biol. 2:139–41.

Pickett, S. T. A., and J. N. Thompson. 1978. Patch dynamics and the design of nature reserves. Biol. Conserv. 13:27–37.

Picman, J. 1988. Experimental study of predation on eggs of ground-nesting birds: Effects of habitat and nest distribution. Condor 90:124–31.

Piergallini, N. H. 1998. Nesting success of the gray catbird *(Dumetella carolinensis)* in a forest irrigated with treated wastewater. Master's thesis, Pennsylvania State Univ., University Park.

Pimental, D., C. Wilson, C. McCullum, R. Huang, P. Dwen, J. Flack, Q. Tran, T. Saltman, and B. Cliff. 1997. Economic and environmental benefits of biodiversity. BioScience 47:747–57.

Pimm, S. L., G. E. Davis, L. Loope, C. T. Roman, T. J. Smith III, and J. T. Tilmant. 1994. Hurricane Andrew. BioScience 44:224–28.

Poole, K. G. 1994. Characteristics of an unharvested lynx population during a snowshoe hare decline. J. Wildl. Manage. 58:608–18.

Porneluzi, P., J. C. Bednarz, L. J. Goodrich, N. Zawada, and J. Hoover. 1993.

Reproductive performance of territorial ovenbirds occupying forest fragments and a contiguous forest in Pennsylvania. Conserv. Biol. 7:618–22.
Pough, F. H., and R. E. Wilson. 1977. Acid precipitation and reproductive success of *Ambystoma* salamanders. Water Air Soil Pollut. 7:307–16.
Powell, D. S., and J. E. Barnard. 1982. Gypsy moth's impact on the timber resource. Pages 72–83 in S. R. Cochran, J. C. Finley, and M. J. Baughman, eds. Proc. of coping with the gypsy moth. Penn State Forestry Issues Conf., Pennsylvania State Univ., University Park.
Powell, D. S., and T. J. Considine, Jr. 1982. An analysis of Pennsylvania's forest resources. USDA–For. Serv., Northeast. For. Exp. Stat., Broomall, Pa. Resourc. Bull. NE-69.
Powell, R. A., and W. S. Brooks. 1981. Small mammal changes in populations following tornado blowdown in northern mixed forests. J. Mammal. 62:397–400.
Powell, R. A., J. W. Zimmerman, D. E. Seaman, and J. F. Gilliam. 1996. Demographic analyses of a hunted black bear population with access to a refuge. Conserv. Biol. 10:224–34.
Prescott-Allen, R., and C. Prescott-Allen. 1990. How many plants feed the world? Conserv. Biol. 4:365–74.
Preston, E. 1973. Computer simulated dynamics of a rabies-controlled fox population. J. Wildl. Manage. 37:501–12.
Probst, J. R., and T. R. Crow. 1991. Integrating biological diversity and resource management. J. Forestry 89(2):12–17.
Probst, J. R., and J. Weinrich. 1993. Relating Kirtland's warbler population to changing landscape composition and structure. Landscape Ecol. 8:257–71.
Pyare, S., J. A. Kent, D. L. Noxon, and M. T. Murphy. 1993. Acorn preference and habitat use in eastern chipmunks. Amer. Midland Nat. 130:173–83.
Rabenold, K. N., P. T. Fauth, B. W. Goodner, J. A. Sadowski, and P. G. Parker. 1998. Response of avian communities to disturbance by an exotic insect in spruce-fir forests of the southern Appalachians. Conserv. Biol. 177–89.
Radcliffe, S. J. 1992. Forestry at the crossroads—integrating economic and social needs with biological concerns. J. Forestry 90(8):22–26.
Ranney, J. W., M. C. Bruner, and J. B. Levenson. 1981. The importance of edge in the structure and dynamics of forest islands. Pages 67–95 in R. L. Burgess and D. M. Sharpe, eds. Forest island dynamics in man-dominated landscapes. Springer-Verlag, New York.
Ratcliffe, D. A. 1967. Decrease in eggshell weight in certain birds of prey. Nature 215:208–10.
Ratti, J. T., and K. P. Reese. 1988. Preliminary test of the ecological trap hypothesis. J. Wildl. Manage. 52:484–91.
Reardon, R. 1995. *Entomophaga maimaiga* in North America: A review. Gypsy Moth News, October:3–4.

Reed, J. M. 1995. Ecosystem management and an avian habitat dilemma. Wildl. Soc. Bull. 23:453–57.

Reeves, R. R., and R. L. Brownell Jr. 1982. Baleen whales. Pages 415–44 in J. A. Chapman and G. A. Feldhammer, eds. Wild mammals of North America. Johns Hopkins Univ. Press, Baltimore.

Regal, P. J. 1977. Ecology and evolution of flowering plant dominance. Science 196:622–29.

Reichman, O. J., and H. R. Pulliam. 1996. The scientific basis for ecosystem management. Ecol. Applications 6:694–96.

Renken, R. B., and E. P. Wiggers. 1993. Habitat characteristics related to pileated woodpecker densities in Missouri. Wilson Bull. 105:77–83.

Ricklefs, R. E. 1990. Ecology. 3d ed. W. H. Freeman and Co., New York.

Robb, J. R., M. S. Cramer, A. R. Parker, and R. P. Urbanek. 1996. Use of tree cavities by fox squirrels and raccoons in Indiana. J. Mammal. 77:1017–27.

Robbins, C. S. 1970. An international standard for a mapping method in bird census work recommended by the International Bird Census Committee. Audubon Field Notes 24:722–26.

———. 1988. Forest fragmentation and its effects on birds. Pages 61–65 in J. E. Johnson, ed. Managing north-central forest for non-timber values. Soc. Amer. For., Publ. 98-04, Bethesda, Md.

Robbins, C. S., D. Bystrak, and P. H. Geissler. 1986. The breeding bird survey: Its first fifteen years, 1965–1979. U.S. Dept. Interior, Fish and Wildl. Serv., Washington, D.C. Resourc. Publ. No. 157.

Robbins, C. S., D. K. Dawson, and B. A. Dowell. 1989. Habitat area requirements of breeding forest birds of the Middle Atlantic states. Wildl. Monogr. 103:1–34.

Robertson, F. D., and R. D. Gale. 1990. Forestry at the national level. Pages 209–30 in R. A. Young and R. L. Giese, eds. Introduction to forest science. John Wiley & Sons, New York.

Robinson, J. G. 1993. The limits to caring: Sustainable living and the loss of biodiversity. Conserv. Biol. 7:20–28.

Robinson, S. K. 1992. Population dynamics of breeding Neotropical migrants in a fragmented Illinois landscape. Pages 408–18 in J. M. Hagan III and D. W. Johnston, eds. Ecology and conservation of Neotropical migrant landbirds. Smithsonian Inst. Press, Washington, D.C.

Robinson, S. K., F. R. Thompson, III, T. M. Donovan, D. R. Whitehead, and J. Faaborg. 1995. Regional forest fragmentation and the nesting success of migratory birds. Science 267:1987–90.

Robinson, W., R. Nowogrodzki, and R. Morse. 1989. The value of honeybees as pollinators of U.S. crops. Amer. Bee J. 129:411–23, 477–87.

Rodenhouse, N. L. 1992. Potential effects of climatic change on a Neotropical migrant landbird. Conserv. Biol. 6:263–72.

Rodewald, A. D. 2000. Influence of landscape composition of forest bird communities. Ph.D. dissertation, Pennsylvania State Univ., University Park.

Rodewald, P. G., and K. G. Smith. 1998. Short-term effects of understory and overstory management on breeding birds in Arkansas oak-hickory forest. J. Wildl. Manage. 62: (in press).

Rogers, L. L. 1980. Inheritance of coat color and changes in pelage coloration in black bears of northeastern Minnesota. J. Mammal. 61:324–27.

Rohrbaugh, R. W., Jr., and R. H. Yahner. 1997. Effects of macrohabitat and microhabitat on nest-box use and nesting success of American kestrels. Wilson Bull. 109:410–23.

Rollfinke, B. F., and R. H. Yahner. 1991. Microhabitat use by wintering birds in an irrigated mixed-oak forest in central Pennsylvania. J. Pennsylvania Acad. Sci. 65:59–64.

Rollfinke, B. F., R. H. Yahner, and J. S. Wakeley. 1990. Effects of forest irrigation on long-term trends in breeding-bird communities. Wilson Bull. 102:264–78.

Romin, L. A., and J. A. Bissonette. 1996. Deer-vehicle collisions: Status of state monitoring activities and mitigation efforts. Wildl. Soc. Bull. 24:276–83.

Rongstad, O. J., and J. R. Tester. 1969. Movements and habitat use by white-tailed deer in Minnesota. J. Wildl. Manage. 33:366–79.

Root, R. B. 1967. The niche exploitation pattern of the blue-gray gnatcatcher. Ecol. Monogr. 37:317–50.

Root, T. L. 1988. Environmental factors associated with avian distributional boundaries. J. Biogeography 15:489–505.

Root, T. L., and S. H. Schneider. 1993. Can large-scale climatic models be linked with multiscale ecological studies? Conserv. Biol. 7:256–70.

Rosenberg, D. K., B. R. Noon, and E. C. Meslow. 1997. Biological corridors: Form, function, and efficacy. BioScience 47:677–87.

Rosenberg, K. V., and J. V. Wells. 1995. Importance of geographic areas to Neotropical migrant birds in the Northeast. Cornell Lab Ornithol., Ithaca, N.Y.

Ross, B. D. 1996. Associations between habitat and bird variables using geographic information systems and bird data. Master's thesis, Pennsylvania State Univ., University Park.

Roth, R. R., and R. K. Johnson. 1993. Long-term dynamics of a wood thrush population breeding in a forest fragment. Auk 110:37–48.

Roy, M. 1995. Unity of voice in endangered species recovery. Conserv. Biol. 9:457–58.

Ruark, G. A., F. C. Thornton, A. E. Tiarks, B. G. Lockaby, A. H. Chappelka, and R. S. Meldahl. 1991. Exposing loblolly pine seedlings to acid precipitation and ozone: Effects on soil rhizosphere chemistry J. Environ. Qual. 20:828–32.

Rudnicky, T. C., and M. L. Hunter Jr. 1993a. Reversing the fragmentation perspective: Effects of clear-cut size on bird species richness in Maine. Ecol. Applications 3:357–66.

———. 1993b. Avian nest predation in clear-cuts, forests, and edges in a forest-dominated landscape. J. Wildl. Manage. 57:358–64.

Rudzinski, D. R., H. B. Graves, A. B. Sargeant, and G. L. Storm. 1982. Behavioral interactions of penned red and arctic foxes. J. Wildl. Manage. 46:877–84.

Rupprecht, C. E., C. A. Hanlon, M. Niezgoda, J. R. Buchanan, D. Diehl, and H. Koprowski. 1993. Recombinant rabies vaccine: Efficacy assessment in free ranging animals. Onderstepoort J. Veterinary Res. 60:463–68.

Rusch, D. H., C. D. Ankney, H. Boyd, J. R. Longcore, F. Montalbano III, J. K. Ringelman, and V. D. Stotts. 1989. Population ecology and harvest of the American black duck: A review. Wildl. Soc. Bull. 17:379–406.

Rusch, D. H., E. C. Meslow, P. D. Doerr, and L. B. Keith. 1972. Response of great horned owl populations to changing prey densities. J. Wildl. Manage. 36:282–96.

Sadinski, W. J., and W. A. Dunson. 1992. A multilevel study of effects of low pH on amphibians of temporary ponds. J. Herpetol. 26:413–22.

Salwasser, H. 1990. Conserving biological diversity: A perspective on scope and approaches. For. Ecol. Manage. 35:79–90.

———. 1991. New perspectives for sustaining diversity in U.S. national forest ecosystems. Conserv. Biol. 5:567–69.

Sample, V. A., and D. C. LeMaster. 1992. Economic effects of northern spotted owl protection—an examination of four studies. J. Forestry 90(8):31–35.

Samson, F. B. 1992. Conserving biological diversity in sustainable ecological systems. Trans. N. Amer. Wildl. and Nat. Resourc. Conf. 57:308–20.

Samson, F. B., and F. L. Knopf. 1982. In search of a diversity ethic for wildlife management. Trans. N. Amer. Wildl. Conf. 47:421–31.

———. 1993. Managing biological diversity. Wildl. Soc. Bull. 21:509–14.

Samuel, D. E., and B. B. Nelson. 1982. Foxes. Pages 475–90 in J. A. Chapman and G. A. Feldhammer, eds. Wild mammals of North America. Johns Hopkins Univ. Press, Baltimore.

Sanderson, G. C., and F. C. Bellrose. 1986. A review of the problem of lead poisoning in waterfowl. Spec. Publ. 4. Illinois Nat. Hist. Survey, Urbana.

Sargeant, A. B. 1972. Red fox spatial characteristics in relation to waterfowl predation. J. Wildl. Manage. 36:225–36.

Sargeant, A. B., and S. H. Allen. 1989. Observed interactions between coyotes and red foxes. J. Mammal. 70:631–33.

Sargeant, A. B., S. H. Allen, and J. O. Hastings. 1987. Spatial relationships between sympatric coyotes and red foxes in North Dakota. J. Wildl. Manage. 51:285–93.

Sauer, J. R., G. W. Pendelton, and B. G. Peterjohn. 1996. Evaluating causes of

population change in North American insectivorous songbirds. Conserv. Biol. 10:465–87.
Saunders, D. A., R. J. Hobbs, and C. R. Margules. 1991. Biological consequences of ecosystem fragmentation: A review. Conserv. Biol. 5:18–32.
Scavia, D., M. Ruggiero, and E. Hawes. 1996. Building a scientific basis for ensuring the vitality and productivity of U.S. ecosystems. Bull. Ecol. Soc. Amer. 77:125–27.
Schacht, A. J. 1993. Nonindustrial forest stewardship. Pages 137–41 in J. C. Finley and S. B. Jones, eds. Penn's Woods—change and challenge. Proc. Penn State Forest Resources Conf., Pennsylvania State Univ., University Park.
Scheel, D., T. L. S. Vincent, and G. N. Cameron. 1996. Global warming and species richness of bats in Texas. Conserv. Biol. 10:452–64.
Schier, G. A. 1975. Deterioration of aspen clones in the Middle Rocky Mountains. USDA Res. Paper INT-170. Intermountain For. Range Exp. Stat., Ogden, Utah.
Schilling, E. A. 1938. Management of the whitetail deer on the Pisgah National Game Preserve (summary of a five-year study). Trans. N. Amer. Wildl. Conf. 3:248–55.
Schindler, D. W. 1988. Effects of acid rain on freshwater ecosystems. Science 239:149–57.
Schonewald-Cox, C., and M. Buechner. 1992. Park protection and public roads. Pages 373–95 in P. L. Fiedler and S. Kjain, eds. Conservation biology. Chapman & Hall, New York.
Schooley, R. L., C. R. McLaughlin, G. J. Matula, Jr., and W. B. Krohn. 1994. Denning chronology of female black bears: Effects of food, weather, and reproduction. J. Mammal. 75:466–77.
Schubert, C. A., R. C. Rosatte, C. D. MacInnes, and T. D. Nudds. 1998. Rabies control: An adaptive management approach. J. Wildl. Manage. 62:622–29.
Schultz, C. B. 1998. Dispersal behavior and its implications for reserve design in a rare Oregon butterfly. Conserv. Biol. 12:284–92.
Schultz, J. C., and I. T. Baldwin. 1982. Oak leaf quality declines in response to defoliation by gypsy moth larvae. Science 217:149–151.
Scott, D. P. 1986. Winter habitat and browse use by snowshoe hares in Allegheny hardwood clear-cut stands. Master's thesis, Pennsylvania State Univ., University Park.
Scott, D. P., and R. H. Yahner. 1989. Winter habitat and browse use by snowshoe hares, *Lepus americanus,* in a marginal habitat in Pennsylvania. Can. Field-Nat. 103:560–63.
Scott, M. J., F. Davis, B. Csuti, R. Noss, B. Butterfield, C. Groves, H. Anderson, S. Caicco, F. D'Erchia, T. C. Edwards Jr., J. Ulliman, and R. G. Wright. 1993. Gap analysis: A geographic approach to protection of biological diversity. Wildl. Monogr. 123:1–41.
Scott, V. E., K. E. Evans, D. R. Patton, and C. P. Stone. 1977. Cavity-nesting

birds of North American forests. USDA–For. Serv., Washington, D.C. Agric. Handbk. 511.

Scribner, K. T., J. E. Evans, S. J. Morreale, M. H. Smith, and J. W. Gibbons. 1986. Genetic divergence among populations of the yellow-bellied slider turtle *(Pseudemys scripta)* separated by aquatic and terrestrial habitats. Copeia 1986:691–700.

Seabrook, W. A., and E. B. Dettmann. 1996. Roads as activity corridors for cane toads in Australia. J. Wildl. Manage. 60:363–368.

Sedjo, R. A. 1990. The global carbon cycle: Are forests the missing link? J. Forestry 88(10):33–34.

———. 1991. Forest resources: Resilient and serviceable. Pages 81–120 in K. D. Frederick and R. A. Sedjo, eds. America's renewable resources: Historical trends and current challenges. Resources for the Future, Washington, D.C.

Servello, F. A., and R. L. Kirkpatrick. 1989. Nutritional value of acorns for ruffed grouse. J. Wildl. Manage. 53:26–29.

Shabecoff, P. 1992. Real Rio: The '92 Earth Summit. Buzzworm 4(5):39–43, 89–101.

Shackelford, C. E., and R. N. Connor. 1997. Woodpecker abundance and habitat use in three forest types in eastern Texas. Wilson Bull. 109:613–629.

Shafer, C. L. 1995. Values and shortcomings of small reserves. BioScience 45:80–88.

Sharov, A., J. Mayo, and D. Leonard. 1998. Results from the gypsy moth Slow the Spread Pilot Project. Gypsy Moth News 44:3–6.

Sharpe, W. E. 1990. Impact of acid precipitation on Pennsylvania's aquatic biota: An overview. Pages 98–107 in J. A. Lynch, E. S. Corbett, and J. W. Grimm, eds. Atmospheric deposition in Pennsylvania: A critical assessment. Environ. Resourc. Res. Inst., Pennsylvania State Univ., University Park.

Sharpe, W. E., V. G. Leibfried, W. G. Kimmel, and D. R. DeWalle. 1987. The relationship of water quality and fish occurrence to soils and geology in an area of high hydrogen and sulfate ion deposition. Water Resourc. Bull. 23:37–45.

Shelford, V. E. 1963. The ecology of North America. Univ. of Illinois Press, Urbana.

Sherwood, S. I., and D. A. Dolske. 1991. Acidic deposition and marble monuments at Gettysburg National Military Park. APT Bull. 23(4):52–57.

Shimanuki, H., N. W. Calderone, and D. A. Knox. 1994. Parasitic mite syndrome: The symptoms. Amer. Bee J. 134:827–29.

Shrauder, P. A. 1984. Pages 331–54 in L. K. Halls, ed. White-tailed deer: Ecology and management. Stackpole Books, Harrisburg, Pa.

Simberloff, D., and J. Cox. 1987. Consequences and costs of conservation corridors. Conserv. Biol. 1:63–71.

Simms, D. A. 1979. North American weasels: Resource utilization and distribution. Can. J. Zool. 57:504–20.

Skelly, J. M. 1989. Forest decline versus tree decline—the pathological considerations. Environ. Monitor. Assess. 12:23–27.

———. 1992a. A word of caution about the causes of forest decline. Amer. Nurseryman. 15 July:95.

———. 1992b. Global perspective on forest health issues: An attempt to discern fact from fiction. Environ. Toxicol. Chem. 11:1049–50.

———. 1993. Diagnostics and air pollution damage appraisals: Are we being sufficiently careful in appraising our forest health? Forstw. Cbl. 112:12–20.

Skelly, J. M., A. H. Chappelka, J. A. Laurence, and T. S. Fredericksen. 1997. Ozone and its known and potential effects on forests in eastern United States. Pages 69–93 in H. Sandermann, A. R. Wellburn, and R. L. Heath, eds. Forest decline and ozone: A comparison of controlled chamber and field experiments. Springer-Verlag, N.Y.

Skelly, J. M., B. I. Chevone, and Y. S. Yang. 1982. Effects of ambient concentrations of air pollutants on vegetation indigenous to the Blue Ridge Mountains of Virginia. Pages 69–82 in R. Hermann and A. I. Johnson, eds. Symposium on Acid Rain: A Water Resources Issue for the 80s. Amer. Water Resourc. Assoc., Bethesda, Md.

Skelly, J. M., D. D. Davis, W. Merrill, and E. A. Cameron, eds. 1989. Diagnosing injury to eastern forest trees. Agric. Information Serv., Pennsylvania State Univ., University Park.

Skelly, J. M., and J. L. Innes. 1994. Waldsterben in the forests of central Europe and eastern North America: Fantasy or reality? Plant Disease 78:1021–32.

Slade, L. M., and E. B. Godfrey. 1982. Wild horses. Pages 1089–98 in J. A. Chapman and G. A. Feldhammer, eds. Wild mammals of North America. Johns Hopkins Univ. Press, Baltimore.

Slocombe, D. S. 1993. Implementing ecosystem-based management. BioScience 43:612–22.

Smith, C. C. 1968. The adaptive nature of social organization in the genus of tree squirrels *Tamiasciurus*. Ecol. Monogr. 38:30–63.

Smith, H. C., N. I. Lamson, and G. W. Miller. 1989. An esthetic alternative to clear-cutting? J. Forestry 87:14–18.

Smith, H. R. 1985. Wildlife and the gypsy moth. Wildl. Soc. Bull. 13:166–74.

———. 1989. Predation: Its influence on population dynamics and adaptive changes in morphology and behavior of the Lymantriidae. Pages 469–88 in W. E. Willner and K. A. McManus, eds. A comparison of features of New and Old World tussock moths. USDA Gen. Tech. Rept. NE-123, Washington, D.C.

Smith, J. S. 1989. Rabies virus epitopic variation: Use in ecologic studies. Advances Virus Res. 36:215–53.

Smith, R. J., and J. M. Schaefer. 1992. Avian characteristics of an urban riparian strip corridor. Wilson Bull. 104:732–38.

Society of American Foresters. 1981. Biological diversity in forest ecosystems. Society of American Foresters, Bethesda, Md.

———. 1992. Choices in silviculture for American forests. J. Forestry 90(2):42–43.

Soulé, M. E., D. T. Bolger, A. C. Alberts, J. Wright, M. Sorice, and S. Hill. 1988. Reconstructed dynamics of rapid extinctions of chaparral-requiring birds in urban habitat island. Conserv. Biol. 2:75–92.

Sovada, M. A., A. B. Sargeant, and J. W. Grier. 1995. Differential effects of coyotes and red foxes on duck nest success. J. Wildl. Manage. 59:1–9.

Speiser, R., and T. Bosakowski. 1988. Nest site preferences of red-tailed hawks in the highlands of southeastern New York and northern New Jersey. J. Field Ornithol. 59:361–68.

Spies, T. A., J. Tappeiner, J. Pojar, and D. Coates. 1991. Trends in ecosystem management at the stand level. Trans. N. Amer. Wildl. Nat. Res. Conf. 56:628–39.

Sprunt, A. 1969. Population decline of the bald eagle in North America. Pages 347–51 in J. J. Hickey, ed. Peregrine falcon populations: Their biology and decline. Univ. of Wisconsin Press, Madison.

Spurr, S. H., and B. V. Barnes. 1980. 3d ed. Forest ecology. John Wiley & Sons, New York.

Steele, M. A., L. Z. Hadj-Chikh, and J. Hazeltine. 1996. Caching and feeding decisions by *Sciurus carolinensis*: Responses to weevil-infested acorns. J. Mammal. 77:305–314.

Steere, A. C., R. L. Grodzicki, A. N. Kornblatt, J. E. Craft, A. G. Barbour, W. Burgdorfer, G. P. Schmid, E. Johnson, and S. E. Malawista. 1983. The spirochetal etiology of Lyme disease. The New England J. Medicine 308:733–40.

Steidl, R. J., C. R. Griffin, L. J. Niles, and K. C. Clark. 1991. Reproductive success and eggshell thinning of a reestablished peregrine falcon population. J. Wildl. Manage. 55:294–99.

Stoddard, H. L. 1931. The bobwhite quail: Its habits, preservation and increase. Charles Scribner's Sons, New York.

Stokes, M. K., and N. A. Slade. 1994. Drought-induced cracks in the soil as refuges for small mammals: An unforeseen consequence of climatic changes. Conserv. Biol. 8:577–80.

Storm, G. L., R. D. Andrews, R. L. Phillips, R. A. Bishop, D. B. Siniff, and J. R. Tester. 1976. Morphology, reproduction, dispersal, and mortality of midwestern red fox populations. Wildl. Monogr. 49:1–82.

Storm, G. L., D. F. Cottam, and R. H. Yahner. 1995. Movements and habitat use by female deer in historic areas at Gettysburg, Pennsylvania. Northeast Wildl. 52:49–57.

Storm, G. L., R. H. Yahner, and E. D. Bellis. 1993. Vertebrate abundance and wildlife habitat suitability near the Palmerton zinc smelters, Pennsylvania. Arch. Environ. Contain. Toxicol. 25:428–37.

Storm, G. L., R. H. Yahner, D. F. Cottam, and G. M. Vecellio. 1989. Population status, movement patterns, habitat use, and impact of white-tailed deer at Gettysburg National Military Park and Eisenhower National Historic Site, Pennsylvania. U.S. Natl. Park Serv. Tech Rept. NPS/MAR/NRTR-89/043.

Storm, H. 1972. Seven arrows. Ballantine Books, New York.

Strauss, C. H., G. L. Storm, R. H. Yahner, and T. Moran. 1989. Expenditure, demographic and attitudinal characteristics of sport hunters using the Delaware Water Gap National Recreation Area. Nat. Park Serv., Tech. Rept. NPS/MAR/NRTR-89/044.

Stribling, H. L., H. R. Smith, and R. H. Yahner. 1990. Bird community response to timber stand improvement and snag retention. N. J. Appl. For. 7:35–38.

Strittholt, J. R., and R. E. J. Boerner. 1995. Applying biodiversity gap analysis in a regional nature reserve design for the edge of Appalachia, Ohio (U.S.A.). Conserv. Biol. 9:1492–1505.

Suarez, A. V., K. S. Pfenning, and S. K. Robinson. 1997. Nesting success of a disturbance-dependent songbird on different kinds of edges. Conserv. Biol. 11:928–35.

Sutcliffe, O. L., and C. D. Thomas. 1996. Open corridors appear to facilitate dispersal by ringlet butterflies *(Aphantopus hyperantus)* between woodland clearings. Conserv. Biol. 10:1359–65.

Svendsen, G. E., and R. H. Yahner. 1979. Habitat preference and utilization by the eastern chipmunk *(Tamias striatus)*. Kirklandia 31:1–14.

Swain, E. B., D. R. Engstrom, M. E. Brigham, T. A. Henning, and P. L. Brezonik. 1992. Increasing rates of atmospheric mercury deposition in midcontinental North America. Science 257:784–87.

Swain, E. B., and D. D. Helwig. 1989. Mercury in fish from northeastern Minnesota lakes: Historical trends, environmental correlates, and potential sources. J. Minnesota Acad. Sci. 55:103–9.

Sweeney, J. M., and J. R. Sweeney. 1982. Feral hog. Pages 1099–1113 in J. A. Chapman and G. A. Feldhammer, eds. Wild mammals of North America. Johns Hopkins Univ. Press, Baltimore.

Swihart, R. K., and R. H. Yahner. 1982. Eastern cottontail use of fragmented farmland habitat. Acta Theriologica 27, 19:257–73.

Swimmer, J., L. Manor, and R. L. Gooch. 1992. Endangered species programs in the 50 states and Puerto Rico. Endangered Species UPDATE 10(2):6–10.

Taylor, G. E., D. W Johnson, and C. P. Andersen. 1994. Air pollution and forest esosytems: A regional to global perspective. Ecol. Applications 4:662–89.

Tear, T. M., J. M. Scott, P. H. Hayward, and B. Griffith. 1995. Recovery plans and the Endangered Species Act: Are criticisms supported by data? Conserv. Biol. 9:182–95.

Temple, S. A., E. G. Bolen, M. E. Soule, P. F. Brussard, H. Salwasser, and J. G. Teer. 1988. What's so new about conservation biology? Trans. N. Amer. Wildl. and Nat. Resour. Conf. 53:609–12.

Terbough, J. 1989. Where have all the birds gone? Princeton University Press, Princeton N.J.

Teulon, D. A. J., T. E. Kolb, E. A. Cameron, L. H. McCormick, and G. A. Hoover. 1993. Pear thrips, *Taeniothrips inconsequens* (Uzel) (Thysanoptera: Thripidae), on sugar maple, *Acer saccharum* Marsh.: A review. Advances in Thysanopterology 4:355–80.

Thomas, J. W. 1990. Wildlife. Pages 175–204 in R. N. Sampson and D. Hair, eds. Natural resources for the 21st century. Island Press, Washington, D.C.

Thomas, J. W., C. Maser, and J. E. Rodiek. 1979. Edges. Pages 48–59 in J. W. Thomas, ed. Wildlife habitats in managed forest: The Blue Mountains of Oregon and Washington. U.S. Dept. Agric., For. Serv., Agric. Handbook No. 553. Washington, D.C.

Thomas, J. W., and H. Salwasser. 1989. Bringing conservation biology into a position of influence in natural resource management. Conserv. Biol. 3:123–27.

Thompson, D. C. 1978. The social system of the grey squirrel. Behaviour 64:305–28.

Thompson, D. Q., and R. H. Smith. 1970. The forest primeval in the Northeast—a great myth? Proc. Annual Tall Timbers Fire Ecology Conf. 10:255–65. Tall Timbers Res. Stat., Tallahassee, Fla.

Thompson, F. R., III. 1993. Simulated responses of a forest-interior bird population to forest management options in central hardwood forests of the United States. Conserv. Biol. 7:325–33.

Thompson, F. R., III, W. D. Dijak, T. G. Kulowiec, and D. A. Hamilton. 1992. Breeding bird populations in Missouri Ozark forests with and without clear-cutting. J. Wildl. Manage. 56:23–30.

Thorne, S. G. 1993. Penn's Woods at the turning point: Creating the future forest. Pages 160–68 in J. C. Finley and S. B. Jones, eds. Penn's Woods—change and challenge. Proc. Penn State Forest Resources Conf., Pennsylvania State Univ., University Park.

Thurber, D. K., W. R. McClain, and R. C. Whitmore. 1994. Indirect effects of gypsy moth defoliation on nest predation. J. Wildl. Manage. 58:493–500

Ticknor, W. D. 1992. A vision for the future. J. Forestry 90(10):41–44.

Tietje, W. D., and R. L. Ruff. 1980. Denning behavior of black bears in boreal forest of Alberta. J. Wildl. Manage. 44:858–70.

Tilghman, N. G. 1989. Impacts of white-tailed deer on forest regeneration in northwestern Pennsylvania. J. Wildl. Manage. 53:524–32.

Tinline, R., C. D. MacInnes, and S. M. Smith. 1994. Measuring the impact of oral vaccination on wildlife. Proc. Ann. Internat. Mtg.: Rabies in the Americas 5:40–41.

Turner, J. W., Jr., J. F. Kirpatrick, and I. K. M. Liu. 1996. Effectiveness, reversibility, and serum antibody titers associated with immunocontraception in captive white-tailed deer. J. Wildl. Manage. 60:45–51.

Tyler, T., W. J. Liss, L. M. Ganio, G. L. Larson, R. Hoffman, E. Deimling, and G. Lomnicky. 1998. Interaction between introduced trout and larval salamanders *(Ambystoma macrodactylum)* in high-elevation lakes. Conserv. Biol. 12:94–105.

Tyser, R. W., and C. A. Worley. 1992. Alien flora in grasslands adjacent to road and trail corridors in Glacier National Park, Montana (USA). Conserv. Biol. 6:253–62.

Ullrey, D. E., W. G. Youatt, H. E. Johnson, L. D. Fay, B. L. Schoepke, W. T. Magee, and K. K. Keahey. 1973. Calcium requirements of weaned white-tailed deer fawns. J. Wildl. Manage. 27:187–94.

U.S. Congress. 1987. Technologies to maintain biological diversity. Office of Technology Assessment, Washington, D.C. OTA-F-330.

U.S. Department of Agriculture. 1968. Forest regions of the United States. U.S. Forest Serv. Map, Washington, D.C.

———. 1976. Southern forestry smoke management guidebook. USDA–For. Serv., Southeastern For. Exp. Stat., Ashville, N. C. Gen. Tech. Rept. SE-10.

———. 1982. An analysis of the timber situation in the United States 1952–2030. USDA–For. Serv., For. Resourc. Rept. 23, Washington, D.C.

———. 1985. Insects of eastern forests. Misc. Publ. 1426. U.S. Forest Serv., Washington, D.C.

———. 1992. Northeastern area forest health report. Rept. No. NA-TP-03B93. USDA–For. Serv., Washington, D.C.

U.S. Department of the Interior. 1989. Summary of public comments on the fire management policy report. USDI, National Park Service, U.S. Dept. Agric., For. Serv., Washington, D.C.

U.S. Department of the Interior. 1995. Federal wildlife fire management policy and program review. Final Rept., USDI–U.S. Dept. Agric., Washington, D.C.

U.S. Department of the Interior. 1996a. Federal wildlife fire management policy and program review. Implementation Action Plan Rept., USDI–U.S. Dept. Agric., Washington, D.C.

———. 1996b. National survey of fishing, hunting, and wildlife-associated recreation. U.S. Dept. Interior, Fish Wildl. Serv. and U.S. Dept. Commerce, Bureau Census. Washington, D.C.

U.S. Fish and Wildlife Service. 1982. The 1980 national survey of fishing, hunting, and wildlife-associated recreation. U.S. Fish and Wildl. Serv., Washington, D.C.

Van Buskirk, J., and R. S. Ostfeld. 1995. Controlling Lyme disease by modifying the density and species composition of tick hosts. Ecol. Applications 5:1133–40.

Vander Haegen, W. M., and R. M. DeGraaf. 1996. Predation on artificial nests in forested riparian buffer strips. J. Wildl. Manage. 60:542–50.
Van Horne, B., and A. Bader. 1990. Diet of nestling winter wrens in relationship to food availability. Condor 92:413–20.
Van Lear, D. H. 1991. Fire and oak regeneration in the southern Appalachians. Pages 15–21 in S. C. Nodvin and T. A. Waldrop, eds. Fire and the environment: Ecological and cultural perspectives. Southeastern For. Exp. Stat., Asheville, N.C. Gen. Tech. Rept. SE-69.
Van Lear, D. H., and T. A. Waldrop. 1989. History, uses, and effects of fire in the Appalachians. USDA–For. Serv., Southeastern For. Exp. Stat., Ashville, N.C. Gen. Tech. Rept. SE-54.
Van Sickle, J., J. P. Baker, H. A. Simonin, B. P. Baldigo, W. A. Kretser, and W. E. Sharpe. 1996. Episodic acidification of small streams in the northeastern United States: Fish mortality in field bioassays. Ecol. Applications 6:408–21.
Vasievich, J. M., D. M. Paananen, C. A. Hyldahl, L. S. Bauer, and W. A. Main. 1993. Training tomorrow's forest resource scientists. J. Forestry 91(3):28–32.
Vaughan, T. A. 1986. Mammalogy. 3d ed. Saunders Publishing Co., New York.
Vecellio, G. M., R. H. Yahner, and G. L. Storm. 1994. Crop damage by deer at Gettysburg Park. Wildl. Soc. Bull. 22:89–93.
Vertucci, F. A., and P. S. Corn. 1996. Evaluation of episodic acidification and amphibian declines in the Rocky Mountains. Ecol. Applications 6:449–57.
Vickery, P. D. 1991. A regional analysis of endangered, threatened, and special-concern birds in the northeastern United States. Trans. Northeast Section Wildl. Soc. 48:1–10.
Villard, M.-A., P. R. Martin, and C. G. Drummond. 1993. Habitat fragmentation and pairing success in the ovenbird *(Seiurus aurocapillus)*. Auk 110:759–68.
Vogelmann, J. E. 1995. Assessment of forest fragmentation in southern New England using remote sensing and geographic information system technology. Conserv. Biol. 9:439–49.
Wake, D. 1991. Declining amphibian populations. Science 253:860.
Walker, B. 1995. Conserving biological diversity through ecosystem resilience. Conserv. Biol. 9:747–52.
Waller, D. M., and W. S. Alverson. 1997. The white-tailed deer: A keystone herbivore. Wildl. Soc. Bull. 25:21–26.
Walter, H. 1973. Vegetation of the earth. Springer-Verlag, New York.
Walthen, W. G., G. F. McCracken, and M. R. Pelton. 1985. Genetic variation in black bears from the Great Smoky Mountains National Park. J. Mammal. 66:564–67.
Wandeler, A. I. 1991. Oral immunization of wildlife. Pages 485–503 in G. M. Baer, ed. The natural history of rabies, CRC Press, Boca Raton, Fla.

Ward, W. W. 1983. Ecological and historical perspective, Pennsylvania oak types. Pages 1–8 in J. Finley, R. S. Cochran, and J. R. Grace, eds. Proc. of regenerating hardwood stands. Penn State Forestry Issues Conf., Pennsylvania State Univ., University Park.

Warren, R. J. 1991. Ecological justification for controlling deer populations in eastern national parks. Trans. N. Amer. Wildl. Nat. Res. Conf. 56:56–66.

Weaver, M. E. 1991. Acid rain and air pollution vs. the buildings and outdoor sculptures of Montréal. APT Bull. 23(4):13–19.

Webb, S. D. 1977. A history of savanna vertebrates in the New World. Part 1: North America. Ann. Rev. Ecol. Syst. 8:355–80.

Webb, W. L., D. F. Behrend, and B. Saisorn. 1977. Effect of logging on songbird populations in a northern hardwood forest. Wildl. Monogr. 55:1–35.

Weber, P. 1994. Safeguarding oceans. Pages 40–60 in L. Starke, ed. State of the world 1994. W. W. Norton and Co., New York.

Webster, P. J., and T. N. Palmer. 1997. The past and the future of El Niño. Nature 390:562–64.

Wegner, J. F., and G. Merriam. 1979. Movements by birds and small mammals between a wood and adjoining farmland habitats. J. Appl. Ecol. 16:349–57.

Weller, M. W. 1989. Plant and water-based dynamics in an east Texas shrub/hardwood bottomland wetland. Wetlands 9:73–88.

Wells, J. V., and M. E. Richmond. 1995. Populations, metapopulations, and species populations: What are they and who should care? Wildl. Soc. Bull. 23:458–62.

Welsh, C. J. E., and W. M. Healy. 1993. Effect of even-aged timber management on bird species diversity and composition in northern hardwoods of New Hampshire. Wildl. Soc. Bull. 21:143–54.

Welty, J. C., and L. Baptista. 1988. The life of birds. 4th ed. Saunders College Publishing Co., New York.

Wemmer, C., R. Rudran, F. Dallmeier, and D. E. Wilson. 1993. Training developing-country nationals is the critical ingredient to conserving global biodiversity. BioScience 43:762–67.

Wheeler, D. L. 1990. Scientists studying "the greenhouse effect" challenge fears of global warming. J. Forestry 88(7):34–36.

Whitaker, J. O., Jr., and S. L. Gummer. 1992. Hibernation of the big brown bat, *Eptesicus fuscus,* in buildings. J. Mammal. 73:312–16.

Whitcomb R. R., C. S. Robbins, J. F. Lynch, B. L. Whitcomb, M. K. Klimkiewicz, and D. Bystrak. 1981. Effects of forest fragmentation on avifauna of the eastern deciduous forest. Pages 125–205 in R. L. Burgess and D. M. Sharpe, eds. Forest island dynamics in man-dominated landscapes. Springer-Verlag, New York.

Whittaker, R. H., and G. E. Likens. 1973. Primary production: The biosphere and man. Human Ecol. 1:357–69.

Wiedenfeld, D. A., L. R. Messick, and F. C. James. 1992. Population trends in sixty-five species of North American birds 1966–1990. Final Rept., Nat. Fish Wildl. Found., Washington, D.C.

Wilcove, D. S. 1985. Nest predation in forest tracts and the decline of migratory songbirds. Ecology 66:1211–14.

———. 1988. Changes in the avifauna of the Great Smoky Mountains: 1947–1983. Wilson Bull. 100:256–71.

Wilcove, D. S., C. H. McLellan, and A. P. Dobson. 1986. Habitat fragmentation in the temperate zone. Pages 237–56 in M. E. Soule, ed. Conservation biology. Sinauer Associates, Sunderland, Mass.

Wilcove, D. S., C. H. McMillan, and K. C. Winston. 1993. What exactly is an endangered species? An analysis of the U.S. endangered species list: 1985–1991. Conserv. Biol. 7:87–93.

Wilcox, B. A., and D. D. Murphy. 1985. Conservation strategy: The effects of fragmentation on extinction. Amer. Nat. 125:879–87.

Wilderness Society, The. 1986. Conserving biological diversity in our national forests. The Wilderness Society, Washington, D.C.

Wildlife Society, The. 1988. Management and conservation of old-growth forests in the U.S. The Wildlifer 227:16.

———. 1992. The role of wildlife management in conserving biodiversity. The Wildlifer 252:21–22.

———. 1993. Conserving biological diversity. The Wildlifer 256:3.

———. 1998. Road policy proposed for National Forest system. The Wildlifer 287:1.

Wildt, D. E., W. F. Rall, J. K. Critser, S. L. Monfort, and U. S. Seal. 1997. Genome resource banks. BioScience 47:689–98.

Williams, B. L., and B. G. Marcot. 1991. Use of biodiversity indicators for analyzing and managing forest landscapes. Trans. N. Amer. Wildl. Nat. Resourc. Conf. 56:613–27.

Williams, C. E. 1994. Forests in decline? Pennsylvania Wildl. 15(6):15–18.

Williams, L. M., and M. C. Brittingham. 1997. Selection of maternity roosts by big brown bats. J. Wildl. Manage. 61:359–68.

Williams, M. 1989. Americans and their forests. Cambridge Univ. Press, New York.

———. 1992. Americans and their forests: A historical geography. Cambridge Univ. Press, New York.

Wilson, E. O. 1988. The current state of biological diversity. Pages 3–18 in E. O. Wilson, ed. Biodiversity. National Academy Press, Washington, D.C.

———. 1989. Threats to biodiversity. Scient. Amer. 261:108–16.

Wilson, M. L. 1986. Reduced abundance of adult *Ixodes dammini* (Acari: Ixodidae) following destruction of vegetation. J. Econ. Entomol. 79:693–96.

Wilson, M. L., and R. D. Deblinger. 1993. Vector management to reduce the risk of Lyme disease. Pages 126–56 in H. S. Ginsberg, ed. Ecology and environmental management of Lyme disease. Rutgers Univ. Press, Piscataway, N.J.

Wink J., S. E. Senner, and L. G. Goodrich. 1987. Food habits of great horned owls in Pennsylvania. Proc. Pennsylvania Acad. Sci. 61:133–37.

With, K. A. 1997. The theory of conservation biology. Conserv. Biol. 11:1436–40.

Wobeser, G., and M. Swift. 1976. Mercury poisoning in a wild mink. J. Wildl. Diseases 12:335–40.

Wolff, G. T., P. J. Lioy, R. E. Meyers, and R. T. Cederwall. 1977. An investigation of long-range transport of ozone across the midwestern and eastern United States. Atmospheric Environ. 11:797–802.

Wood, C. A. 1994. Ecosystem management: Achieving the new land ethic. Renewable Resourc. J. 12:6–12.

Wood, G. W., and R. H. Barrett. 1979. Status of wild pigs in the United States. Wildl. Soc. Bull. 7:237–46.

Wood, P. B., J. P. Duguay, and J. V. Nichols. 1998. Songbird abundance, nest success, and invertebrate biomass on two-age, clear-cut, and unharvested forest stands in the Monongahela National Forest, West Virginia. Final Rept., West Virginia Univ., Morgantown.

Wood, P. B., J. H. White, A. Steffer, J. M. Wood, C. F. Facemire, and H. F. Percival. 1996. Mercury concentrations in tissues of Florida bald eagles. J. Wildl. Manage. 60:178–85.

Wooding, J. B., and T. S. Hardisky. 1992. Denning by black bears in north-central Florida. J. Mammal. 73:895–98.

World Resources Institute. 1988. World resources 1988–89. Basic Books, New York.

———. 1990. World resources 1990–91. Oxford Univ. Press, New York.

Wren, C. D. 1985. Probable case of mercury poisoning in a wild otter, *Lutra canadensis,* in northwestern Ontario. Can. Field-Nat. 99:112–14.

Wyman, R. L. 1990. What's happening to the amphibians? Conserv. Biol. 4:350–52.

Wyman, R. L., and J. Jancola. 1992. Degree and scale of terrestrial acidification and amphibian community structure. J. Herpetol. 26:392–401.

Yahner, R. H. 1977. The adaptive significance of scatter hoarding in the eastern chipmunk. Ohio J. Sci. 75:175–76.

———. 1978. The adaptive nature of the social system and behavior in the eastern chipmunk, *Tamias striatus.* Behav. Ecol. Sociobiol. 3:397–427.

———. 1980. Burrow system use by red squirrels. Amer. Midland Nat. 103:409–11.

———. 1982. Avian nest densities and nest-site selection in farmstead shelterbelts. Wilson Bull. 94:156–74.

———. 1983a. Population dynamics of small mammals in farmstead shelterbelts. J. Mammal. 64:380–86.

———. 1983b. Seasonal dynamics, habitat relationships, and management of avifauna in farmstead shelterbelts. J. Wildl. Manage. 47:85–104.

———. 1983/84. Avian use of nest boxes in Minnesota farmstead shelterbelts. Minnesota Acad. Sci. 49:18–20.

———. 1984. Effects of habitat patchiness created by a ruffed grouse habitat management plan on avian communities. Amer. Midland Nat. 111:409–13.

———. 1985. Effects of forest fragmentation on winter bird abundance in central Pennsylvania. Proc. Pennsylvania Acad. Sci. 59:114–16.

———. 1986a. Structure, seasonal dynamics, and habitat relationships of avian communities in small even-aged forest stands. Wilson Bull. 98:61–82.

———. 1986b. Spatial distribution of white-footed mice *(Peromyscus leucopus)* in fragmented forest stands. Proc. Pennsylvania Acad. Sci. 60:165–66.

———. 1987. Use of even-aged stands by winter and spring bird communities. Wilson Bull. 99:218–32.

———. 1988a. Changes in wildlife communities near edges. Conserv. Biol. 2:333–39.

———. 1988b. Small mammals associated with even-aged aspen and mixed-oak forest stands in central Pennsylvania. J. Pennsylvania Acad. Sci. 62:122–26.

———. 1989. Forest management and featured species of wildlife: Effect on coexisting species. Pages 146–61 in J. Finley and M. C. Brittingham, eds. Proc. of regenerating hardwood stands. School of Forest Resources Issues Symp., Pennsylvania State Univ., University Park.

———. 1990. Wildlife management and conservation biology revisited. Wildl. Soc. Bull. 18:348–50.

———. 1991. Avian nesting ecology in small even-aged stands. J. Wildl. Manage. 55:155–59.

———. 1992. Dynamics of a small mammal community in a fragmented forest. Amer. Midland Nat. 127:381–91.

———. 1993a. Effects of long-term forest clear-cutting on wintering and breeding birds. Wilson Bull. 105:239–55.

———. 1993b. Putting the horse(s) before the cart (biological diversity conservation). The Wildlifer 259:38.

———. 1995a. Forest fragmentation and avian populations in the Northeast: Some regional landscape considerations. Northeast Wildl. 52:93–102.

———. 1995b. Habitat use by wintering and breeding bird communities in relation to edge in an irrigated forest. Wilson Bull. 107:365–71.

———. 1995c. Forest-dividing corridors and Neotropical migrant birds. Conserv. Biol. 9:476–77.

———. 1996. Forest fragmentation, artificial nest studies, and predator abundance. Conserv. Biol. 10:672–73.

———. 1997a. Long-term dynamics of bird communities in a managed forested landscape. Wilson Bull. 109:595–613.

———. 1997b. Habitat fragmentation and habitat loss. Wildl. Soc. Bull. 24:592.

———. 1998. Butterfly and skipper use of nectar sources in forested and agricultural landscapes of Pennsylvania. J. Pennsylvania Acad. Sci. 71:104–08.

Yahner, R. H., W. C. Bramble, and W. R. Byrnes. 1993. Inventorying and long-term monitoring of wildlife and vegetation on the GPU/DQE electric transmission line project. School of Forest Resources, Pennsylvania State Univ., University Park.

Yahner, R. H., and C. G. Mahan. 1996a. Effects of egg type on depredation of artificial ground nests. Wilson Bull. 108:129–136.

———. 1996b. Depredation of artificial ground nests in managed, forested landscapes. Conserv. Biol. 10:285–88.

———. 1997a. Behavioral considerations in fragmented landscapes. Conserv. Biol. 11:569–70.

———. 1997b. Effects of logging roads on depredation of artificial ground nests in a forested landscape. Wildl. Soc. Bull. 25:158–62.

Yahner, R. H., C. G. Mahan, and C. A. DeLong. 1993. Dynamics of depredation on artificial ground nests in habitat managed for ruffed grouse. Wilson Bull. 105:172–79.

Yahner, R. H., T. E. Morrell, and J. S. Rachael. 1989. Effects of edge contrast on depredation of artificial nests. J. Wildl. Manage. 53:1135–38.

Yahner, R. H., J. L. Quinn, and J. W. Grimm. 1985. Effects of a nonpersistent insecticide (ALSYSTIN®) on abundance patterns of breeding forest birds. Bull. Environ. Contamin. Toxicol. 34:68–74.

Yahner, R. H., G. W. Petersen, J. D. Hassinger, R. A. White, G. M. Baumer, B. D. Ross. 1996. Pilot survey of land-use types, rights-of-way, and edges in Pennsylvania using a geographic information system. Final Rept., School of Forest Resources, Pennsylvania State Univ., University Park.

Yahner, R. H., and R. W. Rohrbaugh, Jr. 1996. Grassland birds associated with harrier habitat on reclaimed surface mines. Northeast Wildl. 53:11–18.

———. 1998. A comparison of raptor use of reclaimed surface mines and agricultural habitats in Pennsylvania. J. Raptor Res. 32:178–180.

Yahner, R. H., and B. D. Ross. 1995. Distribution and success of wood thrush nests in a managed forested landscape. Northeast Wildl. 52:1–9.

Yahner, R. H., and D. P. Scott. 1988. Effects of forest fragmentation on depredation of artificial nests. J. Wildl. Manage. 52:158–61.

Yahner, R. H., and H. R. Smith. 1990. Avian community structure and habitat relationships in central Pennsylvania. J. Pennsylvania Acad. Sci. 64:3–7.

———. 1991. Small mammal abundance and habitat relationships on deciduous forested sites with different susceptibility of gypsy moth defoliation. Environ. Manage. 15:113–20.

Yahner, R. H., G. L. Storm, G. S. Keller, and R. W. Rohrbaugh, Jr. 1994. Inventorying and monitoring protocols of vertebrates in national park areas of the eastern United States: The bibliographic report. Tech. Rept. NPS/MAR/NRTR-94/057.

Yahner, R. H., G. L. Storm, R. E. Melton, G. M. Vecellio, and D. F. Cottam. 1992. Floral inventory, cover-type mapping, and wetland inventory of Gettysburg National Military Park and Eisenhower National Historic Site. J. Pennsylvania Acad. Sci. 65:127–33.

Yahner, R. H., and G. E. Svendsen. 1978. Effects of climate on the circannual rhythm of the eastern chipmunk, *Tamias striatus.* J. Mammal. 59:109–17.

Yahner, R. H., and A. L. Wright. 1985. Depredation on artificial ground nests: Effects of edge and plot age. J. Wildl. Manage. 49:508–13.

Young, R. A., and R. L. Giese. 1990. Introduction. Pages 1–4 in R. A. Young and R. L. Giese, eds. Introduction to forest science. John Wiley & Sons, New York.

Zagata, M. D. 1978. Management of non-game wildlife—a need whose time has come. Pages 2–4 in R. M. DeGraaf, ed. Proc. of the workshop for the management of southern forests for nongame birds. USDA, For. Serv. Tech. S. Rept. 1SE-14.

Zak, B. 1965. Aphids feeding on mycorrhizae of Douglas fir. For. Sci. 11:410–11.

Zeitz, P. S., J. C. Butler, J. E. Cheek, M. C. Samuel, J. E. Childs, L. A. Shands, R. E. Turner, R. E. Voorhees, J. Sarisky, P. E. Rollin, T. G. Ksiazek, L. Chapman, S. E. Reef, K. K. Komatsu, C. Dalton, J. W. Krebs, G. O. Maupin, K. Gage, C. M. Sewell, R. F. Breiman, and C. J. Peters. 1995. A case-control study of hantavirus pulmonary syndrome during an outbreak in the southwestern United States. J. Infectious Diseases 171:864–70.

# INDEX

**Richard H. Yahner** is professor of wildlife conservation in the School of Forest Resources and associate dean of the Graduate School at The Pennsylvania State University. Previously, he was chair of the Intercollege Graduate Degree Program in Ecology at Penn State, assistant professor of wildlife at the University of Minnesota, and postdoctoral fellow at the Smithsonian Institution. He has taught in the areas of conservation biology, field wildlife techniques, mammal ecology, wildlife management, and wildlife natural history. He has published more than two hundred research publications on wildlife ecology and conservation biology.